SUSTAINABLE MANUFACTURING

SUSTAINABLE MANUFACTURING

Edited by:

Harinirina Randrianarisoa

ARCLER PRESS

www.arclerpress.com

Sustainable Manufacturing

Harinirina Randrianarisoa

Arcler Press
2010 Winston Park Drive,
2nd Floor
Oakville, ON L6H 5R7
Canada
www.arclerpress.com
Tel: 001-289-291-7705
001-905-616-2116
Fax: 001-289-291-7601
Email: orders@arclereducation.com

ISBN: 978-1-77407-274-5 (Hardcover)

ABOUT THE EDITOR

Hari Randrianarisoa obtained is Ph.D from Institute of Electronics, Microelectronics and Nanotechnology (IEMN) of the University of Science and Technology – Lille (France). His thesis focused for designing and manufacturing of thermal micro sensors. His interests span from on renewable energy especially solar energy. He has a strong knowledge in technology transfer field. He is currently working as a Business advisor for economic development organization in Québec City and as a consultant Scientist advisor.

TABLE OF CONTENTS

LIST OF FIGURES

LIST OF TABLES

LIST OF ABBREVIATIONS

AAIS	accuracy increasing system
AMWU	Australian Manufacturing Workers' Union
ATMP	advanced thermomechanical pulping
BOL	start of life
CDM	clean development mechanism
CHP	combine heat and power
CSR	corporate social responsibility
CSS	capture and storage/sequestration
CTMP	chemo-thermomechanical pulping
DC	direct current
DFE	design for environment
DLAM	distributed linear algebra machine
DR	defect ratio
EOL	end-of-life
EOU	end-of-use
ESP	endless strip production
EU	European Union
HAPs	hazardous air pollutants
HMI	human-machine-interface
LCA	life cycle assessment
LCR	low consistency refining
LIEM	lean implementation and evaluation process
MP	market potential
NNMI	National Network for Manufacturing Innovation
NPD	new item improvement
OECD	Organization for Economic Co-Operation and Development
OECD	Organization for Economic Co-Task and Development

PEG-HCC	polyethylene glycol-hydrophilic carbon clusters
PLM	product lifecycle management
PSS	product-service system
QCM	quartz crystal microbalance
RDF	refuse-derived fuel
RF	radio frequency
RFID	radio-frequency identification
RLs	reverse logistics
RMS	reconfigurable manufacturing system
RMT	reconfigurable machine apparatuses
SD	sustainable development
SDGs	sustainable development goals
SF	streamlined flow
SM	sustainable manufacturing
SMEs	small-sized and medium-sized enterprises
TBL	triple bottom line
UNCSD	United Nations Convention to Combat Desertification
UNESCO	United Nations Educational, Scientific, and Cultural Organization
UNFCCC	United Nations Framework Convention on Climate Change
UNISDR	United Nations International Strategy for Disaster Reduction
VOCs	volatile organic compounds
WBSD	World Business Council for Sustainable Development
WEEE	waste electrical and electronic equipment
WSN	wireless sensor nodes
WSS	wireless sensor systems
WWF	world-wide fund

PREFACE

The Reason for Writing this Book

In the world that is currently being exposed to the increasing amount of pollution and loss of available resources for energy generation and utilization sustainable manufacturing (Sustainable Manufacturing) is a concept that can be very helpful to improve the quality of environment and lives of people.

Sustainable Manufacturing involves the use of renewable resources for energy that put less pressure on the environment and in long term may prove to be very cost effective and healthy for the consumers. In this book, I have tried to lay emphasis on the subject of Sustainable Manufacturing and how it can prove to be a milestone in the existence of the human race.

The Subjects Explored by this Book

The book opens up to the readers by introducing the topic of Sustainable Manufacturing to them. The readers are informed about the value that Sustainable Manufacturing can create for them. The opening chapter of the book also focuses on the various processes of manufacturing and the equipment that can be used to bring sustainability to the process. The corresponding passage also talks about the innovative ways to conserve energy.

The book further moves on to upload the various principles and practices that govern Sustainable Manufacturing. The readers are informed about the various models and methods that have been researched upon so as to achieve the goal of Sustainable Manufacturing. The book talks about the state of art facilities and a global analysis of them and the role that the play in the concerned field.

There is a big importance of Nanotechnology in the development of Sustainable Manufacturing, and thus the book discusses about it in the following sections. It introduces the concept of Nanotechnology to the readers and talks about its importance in the future of Electronics.

It also talks about the importance of Nanotechnology in food processing packaging and distribution the ways in which nanotechnology affects the Aerospace materials are also discussed in following sections. There is also consideration of the use of the nanotechnology in construction sector and medicinal uses. The book further goes on to discuss the challenges that Sustainable Manufacturing can face in the use of Nanotechnology.

The book goes on to dwell upon the challenges and opportunities that the Sustainable Manufacturing faces in different fields like construction medicine electronics et cetera. Moving on there is a discussion about the solution of the challenges mentioned above and the ways in which the processes of Sustainable Manufacturing can be implemented.

The readers are informed about the value that Sustainable Manufacturing can create for them. The consumers the concept of value and sustainable engineering are introduced to the readers.

There is a proper assessment of value creation and the role sustainability play in it also the various patterns that have existed in Sustainable Manufacturing and the share of market they have been able to make has been discussed thoroughly. The various strategies that can come handy in creating value through Sustainable Manufacturing are also discussed with the readers.

There is a very important role that innovation plays in all the major manufacturing processes. In this book, the advances in technology that help Sustainable Manufacturing and promote it are discussed in a proper way.

The use of technology for the development of green manufacturing like use of solar power, laser technology, and other renewable or non-polluting technologies has been focused upon in the concerned sections the areas for development that need priority have also been dwelt upon deeply. The readers are also informed about the impacts that green technology has on their surroundings.

Lastly and most importantly, the future aspects of sustainable development (SD) and manufacturing as discussed in an elaborated manner. It has been envisioned how the world will look in 2050 if it turns itself to Sustainable Manufacturing processes.

The demand of Blue Water by 2050 will increase by many folds and Sustainable Manufacturing can play a vital role in fulfilling this demand. There has also been focus on the greenhouse emissions that can increase to intolerable levels by 2050, and thus the processes of Sustainable Manufacturing prove vital in themselves.

CHAPTER 1

INTRODUCTION TO SUSTAINABLE MANUFACTURING

CONTENTS

The process of sustainable manufacturing (SM) has been defined as the formation of all the products that have been manufactured and which tends to utilize the processes that are being considered as non-polluting, that helps in conserving or preserving the energy along with the various natural resources, as well as are supposed to be sound in terms of economy and thus, are secured for all the pertaining consumers, communities, and the various employees. Differently, it can be said that the process of SM has been providing assistance in evolving the existing technologies to have a conversion of all the materials as well as the products along with the decremented production of the greenhouse gases, lessening up the utilization of the materials that are considered to be non-renewable or toxic, and decreased creation of the waste. The process of SM tends to have an inclusion of the procedure of manufacturing of all the products, which are sustainable (for instance: manufacturing of the energy that is being acknowledged as renewable) as well as the process of SM of all the varied products.

1.1. INTRODUCTION

In the recent time, the terminology of SM is being utilized quite commonly. There pertain so many conferences, which are entirely based on this specified topic, varied reports of the companies, various journal articles and web pages along with the certain claims regarding the provided products or all those activities that have been enhanced so that they could be more of sustainable.

But, many of the professionals or experts have attended the recently conducted NSF workshop so that they could be easily able to have a sound discussion on this topic, and thus, have discovered the paucity or death of the fine description regarding the SM, being considered as one of the greatest or extensive challenges to deal with, which tends to encounter this kind of tract of research.

Whilst there pertains many of the researchers who all are conducting their respective researches on the generic topic based on the sustainability in the various distinct field of study, it appears that not much of efforts are being made to have an application of these outcomes in the process of manufacturing.

All the problems that are being faced in such forms of investigation are being acknowledged as quite challenging as they tend to have an inclusion of the following: valuation, scale, interactions, weighting, aggregation, substitution along with the various responsibilities in order to provide with a name to the few. And, thus, they tend to instantly drive beyond the domain of engineering within such kind of disciplines for example; ecology, moral philosophy, planetary physics as well as economics.

And still, the pertaining manufacturers along with the engineers are being required to take up a few decisions, and for that, they are required to have an adoption of the various operating descriptions till much of the information is being provided related to it.

The pertinent three most generic positions of the fallback are being provided below:

- **Definition:** The Brundtland Commission report (WCED, 1987) has provided with the most extensively utilized description relating to the development of sustainability in the process of manufacturing.
- **Scope:** The generally called Triple Bottom Line (TBL) tends to provide with the normally utilized or adopted scope belonging to the problem or issues related with sustainability in the process of manufacturing that are being utilized by many of the pertaining companies.
- **Evaluations:** Being entirely dependent on the juxtapositions that have been made amongst the various prior or substitute instances, which are making apt utilization of the provided measures related to the normalized performance so that they could be able to provide with the precise estimates or approximates regarding the progress.

In short, some of the observations provided by the researchers have been presented as follows:

- The statement being provided by the Commission of Brundtland has been quite extensively utilized by many of the pertaining manufacturers, various economists along with the engineers, as well as several others, being the operating description regarding the sustainability of the process of manufacturing.

It provides with the definition regarding the development or the evolvement occurring in the sustainability as the growth or the development that tend to attain the requirements possessed by the current generation without sacrificing the potentiality belonging to the future generation in order to achieve their own respective requirements or demands. This provided description is being considered as quite compatible in accordance with the various other elucidations regarding the sustainability of the process of manufacturing (Arrow, 2004). Two of the major facts are that it tends to possess its key focus on their respective, which is considered to be more of a modest objective, and that it tends to have a distinction amongst all the generations, be it the past, present or the future one.

This kind of differentiation distinction, between the requirements of the current generation along with the requirements or the demands of the future generation is being considered as quite essential and thus, will certainly be trying to deviate the key emphasis from what is all is being provided to the pertaining human beings for their well- being, in the recent times (generally it is considered to be a flow of all the products as well as services), to what all is being required to provide with the assistance to the well- being of the human beings in the future which is generally being acknowledged as the process of maintaining or having an accumulation of the varied stocks.

- The generic scope of having sustainability in the process of manufacturing, which is being advised by the TBL, tends to have a recognition of all the three pertaining essential domains that are required to get consideration for having sustainability that generally tends to account for it, having an inclusion of the various domains, be it the economic, environmental or the social one that are frequently being referred to as the pertaining profits, people or the planets.

All such general tracts incompletely agree to the point that have been suggested by the various researchers or the economists that have been provided in here, who have presented with the interpretation of the statements being provided by the Commission of Brundtland so that they could be able to provide with the meaning that the current generation are required to keep up with a specific base of productiveness which are capable of getting acquired by the future generations (Arrow, 2004).

And this specified productive base would be having an inclusion of the three forms of capital, which are: the human capital, the manufactured one along with the natural capital.

Though the systems related to the manufacturing tends to form material wealth for all the respective human beings, they tend to have a utilization or the gobbling up of the extensive quantity of the pertaining resources, during the process of generation or the creation of the great amounts of waste.

The major reason for the conduction of the workshop including the various experts or the professionals, is to transfer or shift the knowledge regarding the state- of- art within the tract of having sustainability in the process of manufacturing, as well as networking amongst the various scientists across the distinct countries so that they could have further progress within this specific arena.

1.2. VALUE CREATION BY SUSTAINABLE MANUFACTURING (SM)

Engineering is exploiting potentials for useful applications.

The process of manufacturing, being the specified discipline in the domain of engineering, generally tends to begin from the thinking as well as imagination possessed by the human beings, from the pertaining information or knowledge regarding the natural phenomena of science, from the creation or formation of the value of the shapes along with the physical materials with the help of the various existing processes within the facet of technology as well as management that have been portrayed in terms of the products, being tangible or intangible, in the provided physical artifacts along with the various services (Figure 1.1).

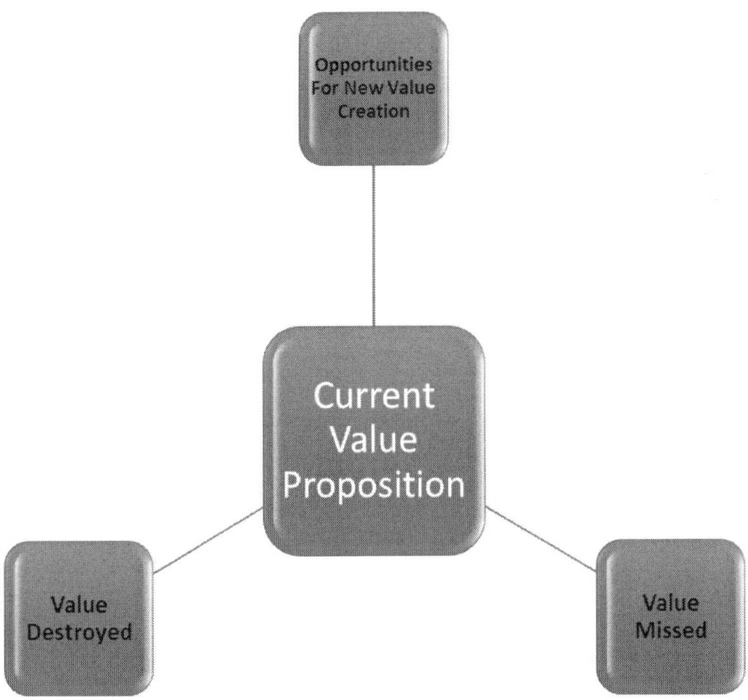

Figure 1.1: Sustainable value creation.

Source: https://link.springer.com/chapter/10.1007/978-3-319-48514-0_13

This research intends to demonstrate how SM embedded in global value creation proves to be superior to traditional paradigms of management and technology.

Having sustenance in the process of manufacturing has become an emergency in terms of the requirement as well as a challenge for the survival or livelihood of the mankind on earth as well as for having development or evolvement for the future, possessing a consideration of the pertinent restrictions regarding the varied resources along with their development or the growth and the uneven dispersion of the provided wealth.

In here, the understanding of the interpretations regarding the terminology of sustainability in the various domains viz: the economic, ecological as well as social one has been presented.

Ecologically, all the pertaining resources which are non- renewable in nature, are required not to get disposed any longer but are supposed to get regained in the various cycles of the material as well as the products.

The likelihood of the replacing all these by the various other renewable are supposed to get exploited, but just to the degree that all such renewable can be capable or possess much of potentiality to get regained.

Teaching and learning for a global culture, wealth, and health become vital tasks for the global human community.

In case of lifestyle of all the forthcoming as well as evolved communities will be provided with certain shapes in the near future, with the assistance of the pertaining, in real the technologies which are predominating, then, in that case, the utilization or the consumption of the varied resources will certainly be exceeding each and every bound, be it social, ecological or economic, that are accountable or credible.

1.2.1. Sustainability Engineering

The sustainable engineering generally helps in presenting a novel approach, being scientific, in order to tackle or deal with the recurring challenges.

Sustainable engineering represents a new scientific approach to cope with this challenge. The changing facets of the competition in the world as well as the collaboration are required to get entirely utilized for all the of global competition and cooperation shall be utilized for providing with an offer regarding wings to the procedures related to the advancement or the innovation and mediation in the direction of sustainability that has been reasonably necessitated across the globe. Thus, a specialized concentration generally tend to pertain on reducing engineering to having sustainability in the process of manufacturing, providing specificity to the process of addressing the generation belonging to the artifacts so that they could be able to provide shape to the livelihood of the human beings.

These two perspectives are merged for methods and tools creating social capital enabling humans for learning and teaching help for self-help (Figure 1.2).

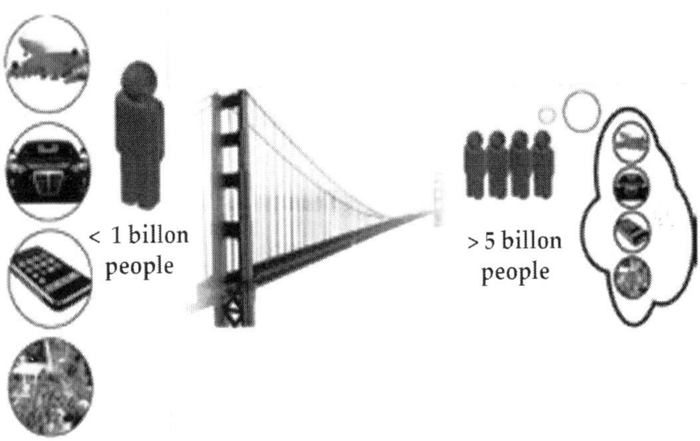

Figure 1.2: From saturated markets bridging the gap to hungry markets.

Source: http://b-ok.cc/dl/2156412/b2dd5a

Though, there pertains various distinction in the lone items belonging to the area of research, the entire concentration generally lies on the process of recognition or the identification of the potentials in various countries for the ingenuities in the process of driving the village, across the globe, regarding the awareness along with the pertaining activities for the development or the evolvement of the sustainability in the process of manufacturing.

Having contributions from the various communities that are on a rise, are required to get identified or recognized for the purpose of having exchange within an collaborative environment along with the various constant advancements or the innovations that have been empowered with the help of the competition as well as the fair trade.

Further cases may specify the implementation of global sustainable value creation in mutual exchange of knowledge between partners from different communities. As it is, being believed that knowledge is the lone resource that has not been lessened up but having an extension by the process of using a robust influence can be assumed in accordance with the various contents related to the manifolds belonging to the management of the pertaining knowledge.

As a result, the various services that are based on the infrastructure belonging to the information being provided or on the generation of awareness amongst the general- public in order to have an exchange of all the perceptions or the opinions along with the stakeholders of the society, as well as a group for providing training regarding the integrated research for all the pertinent doctoral students belonging to the distinctive continents, thus, becoming great ambassadors of the process of having sustainability in manufacturing, within their respective domestic countries along with the various institutions, are considered to be possible.

Whereas, accordance with the perspective being top- down, the various provided networks tend to have an integration of the various modules horizontally as well as vertically, on repeat, on the various distinct levels or stages relating to the aggregation. E.g., original equipment manufacturers "x look for innovative suppliers of physical components or services thus improving their competitiveness.

On the other hand, the varied communities tend to have a glance for the various partners related to academics or education so that they could be able to have an improvement within their social capital along with their respective degree of the wealth.

The modules generally have an inclusion of the equipment, processes, organization along with the various human beings, in the form of the provided factors relation to the formation or the creation of the value.

Valuation of modules in the sustainability perspective is no longer limited to economical but also includes ecological and social criteria.

1.2.2. Complex Challenge of Sustainability

Mankind is often being opposed with the critical challenge related to the issue of sustainability in the process of manufacturing. The basic thought belonging to the development or evolvement of the sustainability has been mentioned within the report of Brundtland in the year 1987.

Formation of the development within the sustainability within the process of manufacturing generally tend to mean that the current generation attains its respective requirements without sacrificing their possessed potentiality of the forthcoming generation in order to achieve their requirements.

In the year 1998, the Commission belonging to the Parliament of various countries provided with their expressions regarding the explanation or the description within their system of three- pillar related to the sustainability being just a perception of the forever development or growth of the sustainability of the process of manufacturing related to the various pertaining dimensions of the economy, ecology as well as the society belonging to the human beings.

In the various researches that have been conducted by the pertaining researchers, approximately 92% of the provided companies have got surveyed and have been provided with the statement regarding the sustainability being the part of their respective strategy of the corporate world.

Mankind has been progressively aware or known of how much they rely on all the process of natural regeneration as well as on the cognizant saving of all those resources, which are considered to be non- renewable.

In the beginning of the 1970s, a report by the Club of Rome "The Limits of Growth" caused international attention on effectiveness and efficiency of resource utilization.

In the recent time, all the scenarios which are being acknowledged as sinister similar to that of a collapse of an ecology that has been created by the arranged growth of the respective population, all the resources that have been exhausted along with the various pertaining risks associated with the pollution have been discussed on a controversial basis.

In the year 2005, a great number of people were found to be greatly influenced or affected by the catastrophes of the weather along with the approximated costs that has incurred of nearly 200 billion US dollar.

A Mankind might be suffering or facing an estimated or predicted cost of approx. 5.5 trillion Euros, in case nothing has been implemented in order to deal with the increasing change or transformations within the climate.

The poorest regions of Africa, Asia, and Latin America will be affected the hardest. All the people living in these tracts would be suffering being under desertification, conversion of the trends or the patterns of precipitation, having deterioration within the production of the agriculture along with the scarcity of water. Various measures related to the index of the human development, which is also being considered as the living standard of the people in the form of the education, health, freedom as well as various other distinct facets of the life that has not being fathomed by the gross GDP. By having a contrast within the utilization or the gobbling up of the various pertaining resources on the several distinct stages of the growth or evolvement along with the eminence of life that are generally being measured in the form of the human development index, wherein, a global increment or progress within the wealth that is entirely based or grounded on the recent advancements in the technology with their respective utilization of the various pertinent resources would certainly be considered as deadly or incurable.

Having comparatively larger strides to the development of the sustainability within the process of manufacturing in the various fragile tracts which are very well defined, and that can be provided with the funds through comparatively rare resource of finances belonging to the world of industries. This results in great opportunities to manage the global challenges of a sustainable development (SD).

Thus, people from all across the globe, are required to have a sound understanding as to how the wealth can be attained or acquired, for anyone, with the help of the various technologies that are considered to be efficacious in terms of the provided resources.

Much of the impartial processes within the education as well as research, formation or creation of the value along with the exchange of the various pertaining products along with the services, are required to get incorporated.

The SM that is being enclosed within the process of formation or creation of the value, all across the globe, are supposed to get proven to be possessing the superiority in relation to the various patterns of the conventional productivity. Value creation is interpreted not only in economical but also in ecological and social dimension.

1.3. MANUFACTURING PROCESSES AND EQUIPMENT

The major objective of having an execution of the various no el techniques along with the tools within the system of manufacturing is to have an increment in the performances of the various novels or the already pertaining process related to the manufacturing.

The tools and techniques that are effective for organizational performance improvements are different in new and old organizations.

Many of the provided proofs tend to provide with the information that some of the presented tools have been found to more of efficacious as well as effective in much of the pertaining institutions or the organizations in comparison to the others.

Some of the researchers have also advised that, whereas the existing norms or the concept that have been provided are found to be identical varying from organizations or the institutions to the other organizations, the real tools that are being utilized in order to prosecute lean are considered to be distinctive.

Thus, many methods, being systematic in nature, are quite essential so as to have a victorious implementation or the execution of the lean, or being triumph in the execution of the various principles or standards related to the process of manufacturing all across the globe, as all these generally tend to possess the roadmaps that provides them with a great illustration of the ongoing status of the respective company with the certain crucial parameters belonging to the performance.

Each metric is developed to evaluate the progress of improvement in a specific area.

1.3.1. A Systematic Approach for Lean Implementation and Evaluation

Many of the researchers have proposed various systematic methodologies in order to have great selection of the opposite strategies related to lean that are entirely based on the various norms belonging to the pertaining institutions or the organization along with the issues or the problems that have been discovered. Thus, the provided figure provides with the flow of the process related to the methodology that has been offered by the researchers (Figure 1.3).

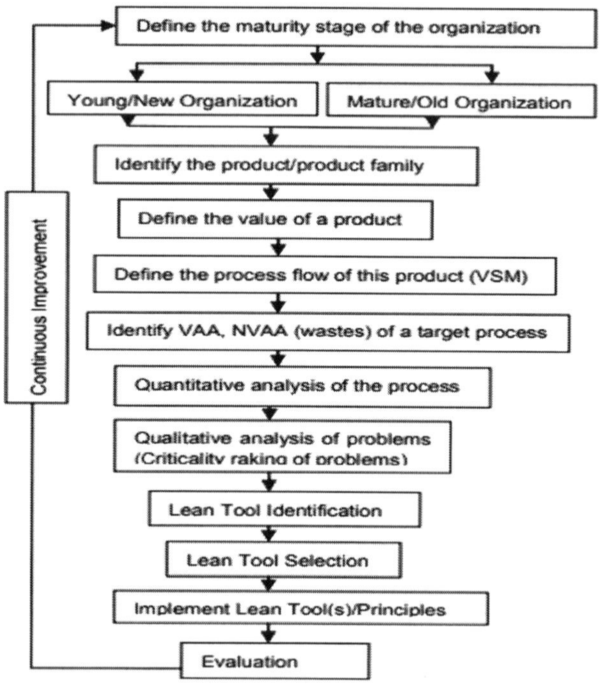

Figure 1.3: A systematic lean implementation and evaluation process (LIEM).

Source: http://b-ok.cc/dl/2098025/ca53da

- **Maturity Stage of an Organization:** Identification of maturity stage (new or old) of an organization is considered as the first step of selecting appropriate lean tools. Certain institutions or the organizations, which are still on their respective early stages related to the life cycle of the organization, generally, tend to have a hold of very less amount of research or the practice on which, they are required to form up their own respective systems of lean.
- **A Product/Product Family:** Even for once, an organization has been discovered or recognized as an organization that is young as well as matured enough, and then it becomes quite essential for them to have an identification of the provided goods or the family belonging to that good so that they could be able to have an enhancement within the process of lean.
- **Value of a Product:** The worth of any of the presented product or good can only be provided with an apt description by the ultimate

consumer or the customer, and thus, it will only be considered as useful, when it will be getting expressed within the terminologies of the presented specified product, which means that a product or the service or at times, both of them will be counted at once, that usually achieves the requirements of the customer at a certain price and at a specified time.

- **Identify the Process Steps (VSM):** In case, the worth or the value of the products or the services have been explicitly defined or provided, then the streams belonging to the value can be capable of getting tacitly discovered.

The process of mapping up of the stream of the value is being considered as the technical methodology belonging to the manufacturing of lean which tends to provide with the assistance to help them interpreting or understanding the provided flow of the various pertaining materials along with the information that is recently being required in order to have a transfer of the products or the services so that they could be able to end the customers.

The lean tools are grouped into three different categories for comparison based on purpose, and each category represents tools that all have similar uses and purposes. Categories are:

- The eminence of the lean or the constant tools for the enhancement;
- Various techniques along with the tools for the process of lean; and
- Certain tools along with the techniques for the support system.

1.4. SUSTAINABLE MANUFACTURING PROCESSES AND EQUIPMENT

The major significance of increasing the life of products has been gradually identified as a part of a quickly shifting agenda of resource efficiency as well as waste management, considerably in reaction to the concerns of environment (Figure 1.4).

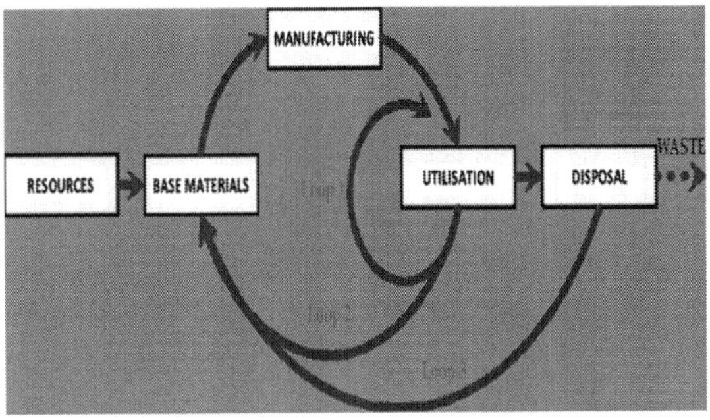

Figure 1.4: The reuse, remanufacturing, and recycling loops.

Source: https://www.researchgate.net/figure/The-reuse-remanufacturing-and-recycling-loops-Adapted-from-Stahel-1982_fig1_266142908

There is well-established body of literature on public values, attitudes, practices, and behavior that fall within this wider agenda, but mainly focusing around the domestic recycling. For the meantime, the developing literature on the notion of remanufacturing comes prodigiously from a perspective which is production oriented.

Then, there is a kind of deceptive gap between the current research that involves with the practices and attitude of people in context with the agenda of material efficiency, and on the other hand, the growing knowledge related to the process of remanufacturing.

Provided that reuse as well as remanufacturing majorly depends on systems establishment which process through numerous production processes, exchange processes, and consumption processes, and back to either exchange or production, this gap is substantial.

1.4.1. Remanufacturing

The concept of remanufacturing is a kind of a considerable element of the UK economy. It is estimated to have approximately a value of £5 billion and to signify UK-wide savings of 270,000 tons of raw materials and 800,000 tons of CO2 (Parker and Butler, 2007).

The development of remanufacturing has prodigiously been in B2B (Business to Business) commodity relationships, with Xerox as well as Caterpillar common exemplars of innovative business methods.

B2C that is Business to Consumer relationships have been very less dynamic in enhancing the notion of remanufacturing. Despite of isolated examples, the process of remanufacturing for end consumers has persisted in the restricted product categories like retreated ink and toner cartridges, tires, and single-use cameras.

Various researches has been done, and it is found that very limited literature, which has any kind of experimental focus on consumers that directly relates to the remanufactured products. Appropriate work has focused on retreaded tires (Fletcher et al., 2003; AEA Technology plc, n.d.); and to single-use cameras and toner/ink cartridges (Lofthouse and Bhamra, 2006).

Fletcher et al., (2003) just after a comprehensive literature research found an absence of work on the perception as well as attitudes of retreads. Lofthouse and Bhamra found that there is very little proof of investigation onto the consumer perceptions of refillable or refill packaging (2006).

A rapidly developing literature on the designing of product for manufacturing comes really closer to get involve with the end consumers, but this review has obtained very less within this literature that has any kind of packaging (Sundin, 2004; Gray and Charter, 2007).

The central framing of consumers in relation with the manufactured goods is to recognize the key challenges faced by the manufacturers in coping with the concerns of end consumers for the fashion as well as status, along with the inevitable negative associations of second-hand goods for consumers (Resource Recovery Forum, 2004; King et al., 2006; Parker, 2007; Parker and Butler, 2007).

1.4.2. Repair

Repaired goods in circuits of re-use attract the smallest specific literature. Generally, repair is a kind of simple process as compared to the process of remanufacturing. Usually, it is understood as a major part of the marginal aspects of the economy.

In the meantime, very little literature defines the public perceptions of the repaired goods as latent purchases. Many researchers in context with the repair has concentrated on the multi-objective social enterprises that either repair or refurbish domestic appliances, classically white goods or computer equipment, mainly for sale to comparatively deprived households (CRR, 2007).

A telephone survey of around one thousand adults in Great Britain asked the respondents regarding the procurement of second-hand or reconditioned electrical appliances. It was found that nearly 2% respondents claim to purchase them every time, whereas 86% have done so on the occasion, barely or sometimes not at all (Brook Lyndhurst, 2004).

That same survey found out that almost 15% assume that purchasing reconditioned or second-hand appliances would make a major difference to the environmental impact, while 515 of the respondents thought, it would not.

It is important to note that the assessments of the gradually increasing efficiency of various appliances, mainly refrigeration and washing appliances defines that rather modern machines are valuable for remanufacturing or repairing on environmental grounds (CRR, 2007).

If a person thinks that a broken product is repairable while someone else might deliberate it as a waste, then it is not an end-of-life (EOL) of that product. The practice, as well as attitude of people to both reuse and repair of their own possessions, is of significance to analyze the attitude of public towards reused and repaired products.

Barr and Gilg (2001, 2005; Barr, 2007) report on an illustrative survey of self-reported recycling, reuse, and reduction behavior and attitudes. Among the queries on reuse, numerous respondents were asked whether they try to repair the damaged things before buying new one.

Around 70% of the respondents said that either usually or always they do so with only 3% claiming that they never did so (Barr et al., 2001).

King et al., (2006) quote a survey that found that around 68% of the respondents cited cost as the main reason why they did not obtained items repaired.

In Japan, research on electric home appliances found that, when respondents wanted to repair and reuse the products, they were conquered by the social system with respondents more often citing the comparative expense of repair when considered against the cost of a new product as an impediment (Tasaki et al., 2004). In both countries, the decreasing comparative cost of purchasing new appliances has counted against repair.

1.4.3. Reuse

As per an ICM telephone survey of the UK house owners for Esure, one in every seven objects in the average UK house is from a second-hand or remanufactured source (Esure, 2006). The survey found out that friends and family are the most extensive source of the second-hand or remanufactured goods, followed by the car boot sales as well as charity shops, with around 3% of respondents taking the discarded products from skips (Esure, 2006).

The survey recognized a bric-a-brac, crockery, glassware, and ornaments, followed by the furniture as the most prospective items to be remanufactured or second-hand, while the electrical goods mainly the white goods to be the least perspective.

As a basic category, reuse incorporates varied phenomenon and a much bigger range of literature related with the practice as well as attitude of the customers in context with the goods falls.

1.4.3.1. Second Hand Exchange

- **Second Hand Retail Outlets:** High-street second-hand trade outlets are the major visible sources of the goods that are second-hand. As per the study led by the Association of Charity Shops, around 69% of the UK population has bought from a charity shop (Association of Charity Shops, 2006).
- Charity shops have been the major point of concern of few consumption-oriented researches related to social science that illustrates the composite variety of values and motives that unite in charity shops and the goods they trade (Gregson et al., 2002), and correspondingly in the characteristic spaces of sovereign retro and vintage clothing outlets.
- **Social Enterprises:** Numerous organizations that provide repaired domestic products and goods, mainly the multi-objective social enterprises are often also conduits of the reused products that do not undergo the repairing process, mainly items of household furniture.
- Few researches have been carried out into the attitudes of those who get furniture by these schemes, along with the motives of the people who contribute to them (Bulkeley et al., 2005; Association of Charity Shops, 2006; Granström, 2006).
- **Online exchange:** Online exchange mainly eBay is of general importance as a part of contemporary secondary markets. Although, the already existing academic literature is conquered by different economists utilizing the unusual form of exchange comprising different mechanisms for public consumer feedback so as to explore dynamics of trust in an extremely mediated

exchange where the traditional bases for trust in a marketing relationship are restricted or sometimes absent (Ba and Pavlou, 2002; Melnik and Alm, 2002; Dellarocas, 2003).

Nissanoff (2006) debates online exchange as a part of a radical vision of existing as well as future changes in how consumers relate to the products and goods. His book is very handy, but it has weedy foundations for its estimates of a consumer economy based on the leasing as well as deliberately temporary ownership of goods based on C2C (consumer to consumer) reuse by online exchange, usually facilitated by commercial mediators.

However, Nissanoff evidently enunciates numerous issues, mainly around trust and ownership, which appear as substantial in considering remanufacturing and reuse.

1.4.3.2. Reuse within an Ongoing Service Relationship

The refillable containers signify an interesting phenomenon of reuse, as they provide insights to the kind of ongoing and closed-loop relationship between producer and consumer that is needed for different progressive models of reuse, servicization, and remanufacturing.

Lofthouse and Bhamra (2006) account a study of attitudes on refillable containers, disclosing the composite factors that unite in the suitability of refill-based services.

From a different kind of perspective, Vaughan, Cook, and Trawick (2007) converse a deep study of the bottle of milk, and the social as well as technical relationships that control its value as a means by which to reflect on the reuse sociology.

1.4.3.3. Reuse of Existing Products and Packaging

Reuse consideration can spread to people are reusing possessions elsewhere their first use, or certainly opting those goods and products that advance themselves to several uses over much more non-reusable alternatives. Barr and Gilg (2001, 2005; Barr, 2007) account on the reuse in accumulation to repair mainly asking different respondents regarding the reuse of paper, jars, plastic bottles, containers as well as reusing the dishcloths.

The majority of respondents appealed that they do these activities either always or so often. The authors do not focus on the consistency of self-reported behavior. The most interesting part from the analysis is the distinct bases for the action of reuse as compared to the bases for the recycling behavior.

Shipton (2005) takes a thorough look at the reuse of packaging, concentrating mainly on the significant uses to which several packaging is put, and the latent for the design to make packaging able to find useful secondary applications, either through designs that enables creativity of people in searching new uses or through designed-in purposes.

Reflecting their growing political currency, relatively there is a huge amount of work on the plastic carrier bags. This is not at all limited in the UK. Ministry of the Environment of Japan reported, "60% of those surveyed have their own shopping bags, showing that the practice of utilizing reusable shopping bags has become popular to some extent."

The survey of Japan also found significantly much more support of around 46.4% as compared to the opposition 28.9% for changing the carrier bags. A study of plastic carrier bag utilization in the UK carried out for WRAP, exposed the complicated relations as well as practices that lie in arrears of the quoted statistics of single-use plastic bag use (Andrew Irving Associates, 2005).

A main finding was that the term "single-use" is something of an inaccurate term with the huge majority of respondents claiming to utilize plastic carrier bags for their subsequent purposes like rubbish disposal or as carrying the sport kit.

By a combination of observation, focused group discussions and omnibus survey, the researchers illustrated how the use, consumption, and reuse of the plastic bags are fixed in the systems, norms, and routines of the shopping environment from the free availability as well as expectations of the use at supermarket checkouts to worries regarding the suspicion of shoplifting if departing a clothes shop with a procurement not in a carrier bag from that shop.

They also demonstrated how something as mundane as a carrier bag is however caught up in the problems of identity, display, and status. An analysis report of a trial bag-for-life (B4L) promotion campaign in the UK found substantial circumstantial determinants of the reuse of carrier bags.

1.4.4. Environment and Waste

The chief driver for the reuse, remanufacturing or political prioritization is the concern related to environment. Proof for how far various concerns over the environmental impact are important in attitudes towards the process of reuse is very different, seemingly depending much on the framing as well as methods of research along with the modes of questioning as on the features of either respondents or the commodities being considered.

In brief, it seems that if individuals are asked regarding the environment, then they will voice potential concern. However, if they are asked that why do they buy reused products, the environment will considerably less prominent, if at all?

Research focusing on the recycling provides different ways into this theme. First of all, the relative normalization of domestic recycling over past few years depicts the different dimensions of altering practices that are relevant to the efficiency of materials.

Until the end of 1990s, surveys related to the recycling behavior of people found that active recycling relates to the concern of an individual for the environment.

With the rapid increase in the facilities of recycling, social norms, as well as expectations of recycling, the process of recycling, has now more likely to be related with the conditional elements like the easy availability of the facility of recycling.

Thus, the recycling dynamics demonstrate how public values, attitudes, and perceptions are dynamic in themselves and have a changing relationship to what actually people do, based upon the systems as well as infrastructure of which they are an essential part.

Current studies have explored the connection between involvement in recycling and the buying of recycled products or the reutilization of the products. It has reliably found gaps that can be taken as following from the de-linking of performing recycling from bigger levels of the environmental commitment.

From the large-scale survey done in Exeter, Barr reports that the reuse of paper, packaging, etc. is forecasted by fundamental environmental knowledge, values, and variables based on concerns, in contrast participating in recycling was considered as highly normative kind of behavior.

In context with the purchase of recycled products, Anstine (2000) reports on a survey of willingness of the consumer to pay for recycled garbage bags in New Jersey, lamenting the absence of evidence for people being prepared to close the loop between their consumption and their active recycling.

From a big-scale survey in Japan, Hanvu et al., (2000) found a same disjuncture between recycling and purchasing recycled, with the behavior of recycling being defined by the system of waste collection in place along with different payment systems, while willingness to purchase recycled toilet paper was linked with the individual environmental attitudes.

Coming from an edging of research in context with the resource efficiency or sustainability, reuse behavior looks strongly linked with the values of environment. Although a very distinct image evolves from research, it seems at the buying of second-hand or remanufactured goods and involvement in different kinds of second-hand utilization.

For instance, the Association of Charity Shops Survey found that just 5–6% of donors as well as buyers denoted to matters related to environment at all as a reason for buying from charity shops, concluding that few people seem to link reuse to environmental values (Association of Charity Shops, 2006, p. 2).

Gregson and Crewe (2003) converse how they started their fieldwork on second hand or remanufactured goods consumption expecting to come across the talk of green consumerism as well as commitment to maximizing the value of resources, at least among critical middle-class respondents.

While they did come across green and environmental concerns in context to the purchase of second-hand goods, it was a smaller part of more complicated accounts of why people purchase second-hand goods.

1.4.5. Price

Naturally, the fact that second-hand or remanufactured reused, repaired goods are quite cheap is the leading issue in mainstream accounts of why people purchase them. The importance of price is borne out recurrently, but playing out in a different way as per the multiple dimensions that can join in different situations of reuse.

In remanufactured goods, the dynamics of markets clearly diverge depending upon their comparative price benefit in comparison with modern goods. Debo (2005) recommends that remanufactured or second-hand goods have basically less value for consumers as compared to new.

Therefore, universal industrial restructuring, as well as the real-terms reducing the cost of products (mainly quite complex technologies that classically advance themselves to remanufacturing), has totaled against the development of remanufacturing (Resource Recovery Forum, 2004; CRR, 2007; Parker and Butler, 2007). This is borne out too in context to the retreaded tires, where the historic price benefit of retreads has been damaged by the collapse in modern tire prices. Prospects of low price of remanufactured as well as reused products conveys through to refillable packaging with low comparative price of refills, as against new products (Lofthouse and Bhamra, 2006).

In the same way, a study of business models for the product service system recommended a key reason for consumers purchasing into services systems like pay-per-use is to access products that the consumer could not anyway afford (Adams et al., 2005).

1.5. PRODUCT DESIGN FOR RESOURCE EFFICIENCY AND EFFECTIVENESS

The last hundred years of the industrialized society have been conquered by basically a straight way of production as well as consumption in which the raw materials are utilized so as to manufacture goods and materials that are sold and then used, and ultimately discarded as waste. Although, most of the resources needed for making a product be on a restricted scale (Figure 1.5).

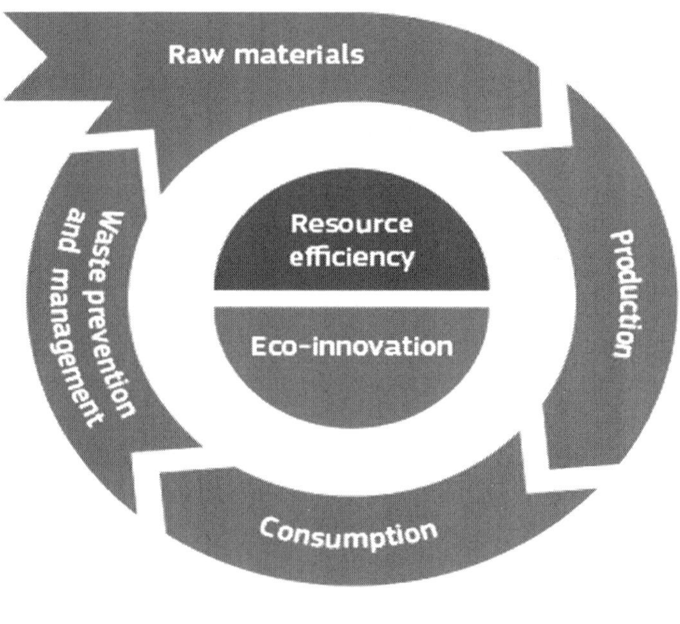

Figure 1.5: Product design for resource efficiency.

Source: http://ec.europa.eu/environment/green-growth/resource-efficiency/index_en.htm

This mainly applies to the raw materials and input goods, mainly those that are non-renewable. In international context, water, and land are also a kind of limited resources. Whilst energy is not essentially a limited resource per se, the utilization of fossil fuels requires to be reduced radically, due to emissions of greenhouse gases from its production as well as use.

Biofuels may also be considered as a resource, which is limited while considering the water and land that are used to produce them. Additionally, energy is utilized not only when goods are processed but generally when they are utilized in their EOL phase.

The pursuit of a significant improvement in utilization of resource and performance has led the society comprising business, to look for better methods of utilizing and then reusing products and their components, and their components, and also to restore more of the energy, material, and labor inputs.

This means that huge demands are gradually imposed on several industries for product designing that, not only in the phase of production but also during the whole life cycle, is as resource-efficient as well as energy-saving as possible.

European provision of input goods comprising raw materials is discussed in the European Innovation Partnership. There all of the representatives of the research, business sector and politics are comprised in a high-level group. Sweden gets involved in this cooperation and a strategic execution plan citing proposed measures has been drawn up.

As there are numerous actors in Sweden, both in the academic sectors and the business sectors that are active in the wider field of resource efficiency, there is yet no coherent program that guise upon the issue from a wide perspective.

It is usual that the field extends over different academic disciplines, indicating that an interdisciplinary method comprising different disciplines as well as stakeholders are required.

For instance, engineering research on the notion of remanufacturing, research on the designing of new products, which are further optimized for maintenance, remanufacturing, and reuse, the relationship between services and products, and the understanding of the models of new business along with the economic aspects on micro as well as macro levels, must be dealt with in the similar program.

Considering the greater dependence of the economy of different countries on the industrial sector and the fruitful as well as long cooperation between academia and business, it is believed that the proposed program is required, has the power to deliver and can easily make a major contribution to the development of the field globally.

On the global level, the UNEP that is the UN Environment Program has developed its International Resource Panel in the year 2007 so as to provide coherent, independent, and authoritative scientific assessment on the sustainable utilization of the natural resources as well as the environmental impacts of the resource use over the complete life cycle.

1.5.1. Design for New Products

It is required that the design for advanced life-long products increases the lifetime and also increases the time beforehand a product requires repair and maintenance. The repair and maintenance methods must only need very little amount of energy, raw material, and labor.

The repair and maintenance system may be designed so that the damaged product is yet working properly during the phase of repair and maintenance.

Additionally, the design must consider those resources that are used when the product is active, contra passive utilization like consumption of energy during user mode as well as sleep mode. Also, it is relevant to study as well as influence customer patterns.

For instance, the framing and designing of the products having short-life may encourage consumers to utilize refill systems or sometimes enable them to completely use the product with less of waste.

In case of long-life products, the design may encourage the consumer to turn off the energy supply when the product is not in the phase of use, have a significant consumer pattern in use, increase the lifetime for the product because of correct handling, etc.

Additionally, the design may encourage consumption of energy during transportation as well as resources required for handling and storage at the wholesaler.

Developing as well as implementing tools like Life Cycle Assessment (LCA), ISO systems, eco-design etc. that determine and enable a systematic resource effective design of new products in the view of raw material consumption, energy as well as labor during construction, utilization, and at the end of product life is also within the scope of this program. In various cases, the tools previously exist, but their usability needs to be enhanced so that they are executed and easily used in a mature way. Understanding the utilization of eco-design, related tools and the practice in a context of company comprising the trade-offs that these tools certainly give rise to, is a significant aspect to consider.

1.5.2. Remanufacturing and Technology Updating

Remanufacturing of old as well as new mechanical equipment like automotive, aeronautic, production, and rolling stock, building equipment, pumps, bearings, robots, electronic, and electrical systems or products, medical equipment comprises various issues and challenges that are in the scope.

It comprises social behavior and the improvement of the public awareness of the quality as well as possibilities of remanufactured goods. So as to enhance the use of remanufactured goods by improving the policy strategies such as public procurement, new business models have to be developed, and new value chains have to be found out.

Analysis strategies, when it is important to remanufacture at all, in view that few old products may have a negative impact on the environment as compared to new products, execution of analytic tools like LCA that reveal the profit as well as losses with the manufactured products, are also involved.

There are various interesting and practical issues related to the improvement of existing processes of remanufacturing, along with the creation of new ones. These consist of, tackling the issue of forecasting the situation and the quantity of incoming goods, to improve the assembly or disassembly processes, to develop new mechanical tools, to enhance the process of cleaning, etc.

1.5.3. New Business Models

Developing as well as recognizing the new business models is completely in the scope of the SM. The models of new business along with the accompanying product, services, and manufacturing system may, for example, concentrate on the solution and performance instead focusing on the product itself. The built-up competence will be very much essential for the chance of creating new markets.

Examples of the conditions where different business models would encourage the resource efficiency comprise those models that ensure as well as promote use-intensity of the system and products, decreasing the redundancy while at the same point of time giving attractive application solution to the consumer, like increasing the diversity of solutions and reducing the burden of maintenance.

For such kind of strategies of the product or service system to become successful, both the business model and the product required to be designed in such a way that it permits for the complete potential to be realized.

Additionally, it will be necessary in few scenarios so as to assure that policy as well as legislation does not delay the efficient solution that break out the existing linear economic model of product consumption, for example, product, taxes, liability, etc.

1.6. INNOVATIVE ENERGY CONVERSION

The various systems related to the products and service provides with an offer regarding the chances or the opportunities for the pertaining companies, which all are keen in having a great benefit within the place of market, in today's time by presenting with an introduction of the certain novel ways of tackling with the value chain of the existing customers along with the various businesses.

In accordance with such a sense, it is required to get considered that the major involvement of the model belonging to the business is to possess a sound understanding of it, know about their respective structure or design, having a great analysis as well as having a transformation within their logics related to the existing organizations.

However, the recent literature tends to present with an offer of the slight information regarding the pertinent model of the respective business, which is considered to be quite apt for the approach of PSS, therefore, having enough to afford for an opportunity or the chance for a sound analysis.

Finally, an exploratory case study of a machine tool manufacturer was performed to identify characteristics for the definition of its business model based on PSS approach.

1.6.1. Product-Service System (PSS)

The PSS generally tends to have an involvement of the various products that are tangible along with the services that are known to be intangible, which have been designed as well as amalgamated so that they could be easily collaboratively capable of satisfying the requirements possessed by the various consumers or the customers.

In case of the conventional models related to business, the customer often buys a product and thus, become quite accountable so that he/ she could be able to explicitly analyze their respective performances, having sustenance as well as sufficient disposal.

In accordance with such a distinction as well as the norms set by PSS, the provided manufacturer tends to earn much of profits from the customers or the consumers for the process of presenting them with the function.

1.6.2. Business Model

Shafer et al. has provided with a description regarding the concept of the model related to the business, being the depiction of the main logic belonging to the company. Therefore, it is being acknowledged as a robust tool or technique, which will be providing them with the assistance to have a sound communication of their respective choices relating to the strategies made.

Hence, it gives the organization a holistic view of the business, clarifying important issues and relationships. The business model can be also considered an abstract representation of the business strategy.

Thus, in accordance with the in-depth analysis of the various researches that have been conducted based on the concepts or the norms related to business model, along with its attributes as well as the association it has, Osterwalder et al. has provided with the description regarding the business model, being a tool based on the various concepts is usually made up of various characteristics, objects along with their relationships or the connections, which generally presents with a shortened explanation regarding the business logic of the company.

REFERENCES

1. Davim, J., (2018). *Sustainable Manufacturing*, p. 252.
2. Ijomah, W. C., McAloone, T. R., Stahel, W., Sundgren, J., & Ahlqvist, J., (2014). *Product Design for Resource Efficiency* [eBook] (p. 22). Retrieved from: https://www.mistra.org/wp-content/uploads/2018/01/Product-design-for-resource-efficiency.pdf (Accessed on 29 March 2019).
3. *Introduction to Sustainable Manufacturing*. [eBook] (p. 2). Retrieved from: http://www.gian.iitkgp.ac.in/files/brochures/BR1467432445Brochure__GSD_Rao_Intriduction_To_Sustainable_Manufacturing.pdf (Accessed on 29 March 2019).
4. Seliger, G., (2012). *Sustainable Manufacturing Shaping: Global Value Creation*, p. 383.
5. *Sustainable Manufacturing*, (2014). [eBook] (p. 5). Retrieved from: http://web.mit.edu/2.813/www/Projects/ProjectBackground2014.pdf (Accessed on 29 March 2019).
6. *United Nations Conference on Trade and Development*, (2015). [eBook] (p. 19). Retrieved from: https://unctad.org/meetings/en/SessionalDocuments/ciid29_en.pdf (Accessed on 29 March 2019).
7. Watson, M., (2008). *Public Attitudes, Perceptions, and Behavior Relating to Remanufactured, Repaired and Reused Products.* [eBook] (p. 30). Retrieved from: http://www.remanufacturing.org.uk/pdf/story/1p322.pdf (Accessed on 29 March 2019).

CHAPTER 2

MODELING AND TACTICS FOR SUSTAINABLE MANUFACTURING

CONTENTS

The practices being related with the Sustainable manufacturing (SM) that has been identified by various pertaining organizations are considered to be the core components in the process of having an increment within the performance as well as competencies of the respective businesses. Whereas all the practices, which have been reported are being acknowledged as great instances related to what all has been attained, they are generally specified to the companies and thus are known for being quite critical for the other people to reproduce as they tend to present with very less, in case any, details regarding the persistent progress as well as developments and how they are being achieved. Various strategies relating to the SM usually provide with an insight about the entire approach, which is required, by all companies or the organizations but they might possess a dearth or paucity of the pragmatic assistance for the execution. Thus, this chapter provides with the assessment of the existing gap between the direction or the way of the strategies along with the certain practices which are included in the process of extraction of the procedures behind such practices and therefore, helps in the formulation of the tacts related to the SM (that helps in presenting with the required information which are based on how to execute all the specified enhancements).

2.1. INTRODUCTION

The process of manufacturing has always been conventionally linked or connected with the various side effects side effects being related to the environment and are unwanted because the manufacturers are being considered as responsible for the process of conversion of such resource inputs into the outputs which are of great use to them, i.e., traditionally products possessing value in terms of economy along with the pertaining restrictions on the efficaciousness because of the laws belonging to the thermodynamics.

Since past forty years, the pressure or stress, belonging to the environment has been associated with the activities performed by the various industries, has witnessed a rise in its essentiality in the issues pertaining in the world along with the extensive challenge for the society in which one lives.

Sound understanding or awareness regarding the effect of activities relating to the human beings on the global environment has tend to have an increment in the practices that exists for the aversion or avoidance of the degradation of the respective environment. Practices similar to these can be discovered easily under certain fields as well as labels, for instance,

Management of the Life Cycle of Product, Ecology of various Industries, Management of Corporate Environment, Management of the Green Supply-Chain as well as the Design for Environment (DFE), Systems for the Product-Service, and various others.

There pertains umpteen factors which are playing a considerable part in the process of providing a definition to all the pertinent need or requirement for the population belonging to the next-generation having a standard for manufacturing, for instance, the grown criticalness or the complexity of the pertaining systems as well as the product, the concerns relating with the environment, paucity or dearth of required integration of knowledge, various innovations along with the advancements in technology involved within the techniques belonging to the process of simulation and modeling.

Quite lately, the main abstraction belonging to the SM has got established being under certain labels (for instance: the process of manufacturing in the environment staying conscious or the process of green manufacturing) being the only sub- norm relating to the avoidance or the aversion of pollution.

One of the key targets or the goals of SM are to reduce the influence or the effect on the environment that has been associated with the process of manufacturing. Various activities, which are being related to the environment, have been connected with the pessimist influence or effect on the performance of the business since very long but the pertaining researchers have been successful in providing evidence regarding the falsity of all such assumptions associated.

A description regarding both the benefits relating to the economy as well as the environment of SM is quite implied in the process of saving up the cost because of the decrement or reduction within the required energy along with the minimization of pertaining waste.

The field belonging to the SM related to the research is progressively developing, and thus, there exists no incorporated description or definitions or outliners for studying the performance of sustainability by the pertinent systems regarding the process of manufacturing.

For the entire literature, the courses relating to the various resources in the form of material, energy along with the wastes that have been connected with it (MEW) might tend to appear again.

The flows belonging to the Material, Energy as well as associated Wastes resource are required to get understood in the broadest of the forms so that they could be able to have an inclusion of not only the process of transforming the first and foremost material but also the various other relevant inputs along with the wastes for instance: all the packaging, expendables as well as water.

SM can be considered as being the strategy relating to the process of manufacturing, which provides with assistance in integrating certain acknowledgments relating to the environment as well as society, being an addition to the acknowledgments of economy along with the technology.

The entire information that has been provided within this chapter tends to possess the main focus on the varied paces or facets of environment as well as having an emphasis on the on-site solutions instead of the life cycle of the product or its chain of supply.

Being specific, the work generally tend to focus on the various common tacts so that they could be able to have an improvement in the flows related to the Material, Energy along with the associated Wastes within the pertaining system related to the process of manufacturing and provides with an approach with the help of which it could get assessed or evaluated.

Such tacts have been created by the process of extraction of the procedure behind the practices related to the SM that has been collected by the various researchers and got formulated so that they could be able to get applied widely across varied resources along with the technologies.

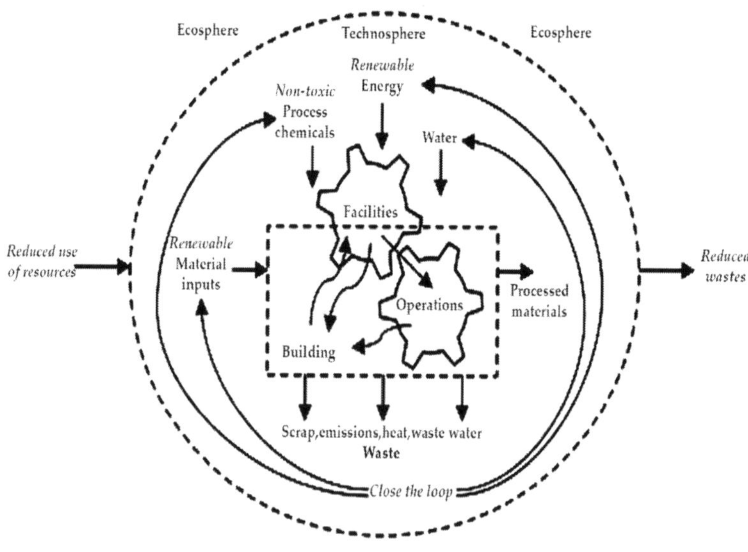

Figure 2.1: Manufacturing (eco) system model with the sub-systems and resource flows.

Source: https://www.researchgate.net/publication/323253652_Sustainable_Manufacturing_Tactics_and_Modelling_An_Improvement_Methodology_for_Manufacturers/download

This wholly means that all such tacts are required to be common or general so that they can perceive or apprehend the pertaining principles regarding the enhancements, but are being provided with enough details so as to get modified in accordance with the meticulosity of the system that has been researched on by the various researchers (Figure 2.1).

The research that has been conducted is being based on great analysis as well as the collection regarding the case studies that are accessible in consideration with the interaction with pertaining industries relating to manufacturing.

The amalgamated utilization of resource flow (for instance: material, energy along with the associated wastes), varied tacts along with the modeling are capable of providing with the aid to all the manufacturers in their respective voyage towards sustainability of the process of manufacturing by presenting with the common solutions regarding how to grasp their provided functions or operations.

Thus, the enhancement within the methodology has been evolved by combining the model relating the ecosystem of manufacturing along with the pertinent tacts so that they could provide with the required guidance to all the manufacturers within a well- defined or structured as well as a systematic way in order to have an identification or recognition of opportunities or chances relating with the developments.

This chapter thus have explored the various structured challenges relating with the development of the methodology belonging to having an improvement within the process of manufacturing so that they could be able to provide assistance to the pertaining users in the process of recognizing regarding what all tacts may get applied within this specified norms.

Utilization of the model relating to the ecosystem of manufacturing, various techniques regarding the modeling techniques, which can apprehend the flows, related to the Material, Energy along with the associated wastes with the help of a system related to manufacturing.

Thus, it generally takes the user with the help of the methodologies that have been improved in order to have an identification of the various pertaining opportunities for the purpose of having improvement within the productivity of the various resources having a great utilization of the common or general tacts so as to shift in the direction of SM.

2.2. EXPLORING SUSTAINABLE MANUFACTURING PRINCIPLES AND PRACTICES (FIGURE 2.2)

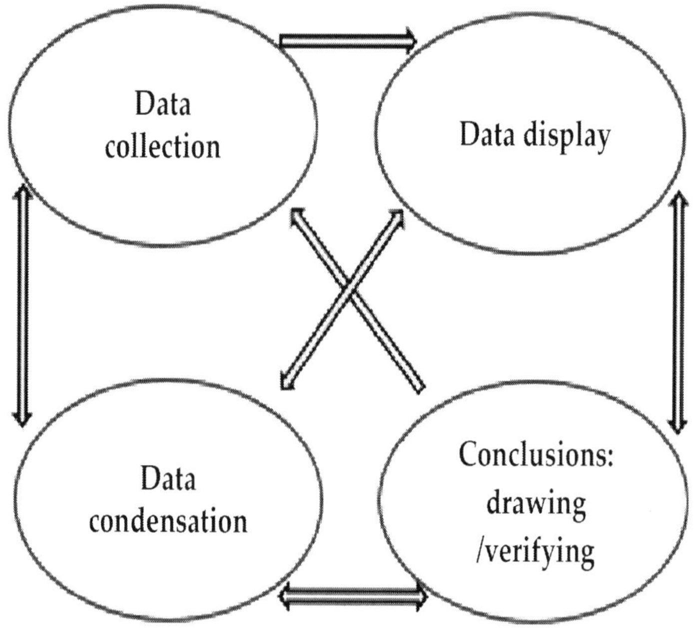

Figure 2.2: Exploring sustainable manufacturing.

Source: https://www.semanticscholar.org/paper/Exploring-Sustainable-Manufacturing-Principles-and-Alay%C3%B3n/614df1b0c1896925842379f2f4b818efbac7c4ad

2.2.1. Sustainability in Manufacturing Operations

All the existing companies have always been acknowledged as the main players who are on the path of society in the direction of maintenance or sustainability (Koplin et al., 2007, p. 1060).

Even in the case of ancient times, many of the pertaining companies have segregated the various acknowledgments relating to the sustainability from their respective strategies or planning related to business along with the assessment or evaluation of the various performances; thus, in accordance with such a way, the pertinent indicators belonging to the performance of the economy have always been considered as the key norms with the help of which the companies are being evaluated (Clarkson, 1995).

This scenario began to change in the late 1980s when a few manufacturing sectors started adopting practices to help with the minimization or elimination of waste. Thus, all these were being considered as the primary indicators of the SM (Lazlo et al., 2013; Courtice, 2013).

The rising pattern or trend regarding the consideration of sustainability of settings within the process manufacturing has generally been quite explicit in the initiatives taken up by the government, for instance: the strategies related to the growth of various countries across the world that usually asks for the combined efforts towards the countries being efficacious in terms of resources.

Such an initiative by the government of various countries has been targeted towards providing commitment to both the Union of the respective country along with the members of states in consideration with the agenda of sustainability, providing with an aid to the occurrence of shift in it towards the economy possessing low carbon, and thus, doing the endorsement of the surge in the utilization of the various sources belonging to the renewable energy, without sacrificing the efficaciousness of the energy.

Nonetheless, in the recent times, it is not just the makers of the set policy who generally tend to follow the incorporation of the society which is sustainable in terms of economy, environment as well as the society, but also all those consumers who possess sound understanding or awareness regarding the problems or issues related with the sustainability all along with the provided stages of the life- cycle of a pertinent product that has witnessed a considerable rise (Vermeir and Verbeke, 2006; UNEP, 2012; Nielsen; 2015).

All such conditions, which are considered to be demanding, have tend to motivate many of the manufacturers to provide with revert to the issues related with sustainability. Global economies have robustly been affected or impacted by the sector belonging to the process of manufacturing.

Such a sector has been created or made up with the help of an extensive variety of the existing subsectors that ranges from basic to advanced practices along with the various provided techniques in the process of production, having an involvement of distinct sizes of the firms, presenting with the generation of growth or development as well as the facilities related to employment. Various datum along with the reports that has been presented by the statistical segregation of activities related to the various economies in terms of statistics in the certain communities (NACE), which has provided with the position of second-leading sector of economy in the domain of manufacturing in the union of various countries, having an approximation of millions of companies that are employing hundreds of millions of people and thus are helping in having a contribution of nearly 22.4% of employment within the Union of various countries within the economy belonging to the nonfinancial businesses economy (Eurostat, 2012).

These statistics reflect how the industrial sector plays a significant role in the economic well-being of Europe (European Commission, 2015). However, this kind of sector has also been held responsible for great pessimist effects on the sustainability of the environment across the globe.

Being an example of the main part the pertaining sector tends to play within the process of sustainability of the global, the sector belonging to the industry generally tend to have the utilization or consumption of the similar expanse of energy and thus, tends to exhibit the similar kind of greenhouse gases being the sectors belonging to the construction along with the transportation (IEA, 2012; UNIDO, 2014).

Still, the energy demands of the industrial sector have been growing faster than in any other sector, contrasting with the continuing efficiency gains (IEA, 2012).

One of the majorly essential events, which has provided with the assistance in order to incorporate as well as promulgate the pertaining norms or concept related with the sustainability was considered to be the issuance of "Our Common Future" (Brundtland, 1987), under which the primary description or the definition relating to the growth or development of sustenance has been presented.

2.2.2. Sustainability Principles

In accordance with the diction providing with the description regarding the terminology of principles, which tends to refer to a great set of policies that has already been fixed or been determined in advance, various ways relating to the action or the provided manner.

The description being provided by Glavič and Lukman, regarding the principles states that these are being considered as the ultimate norms or the concepts, which tend to operate as the base for all the actions that are being performed, and being an important context for the purpose of incorporating a system that is required to be more of complex. These are some of the many definitions where the term "principles" alludes to guidance for further work. Similarly, principles have been defined as "the key concepts that serve as a basis for action."

Daly (1990) was being acknowledged as the first one to have made an attempt towards the publishing of various principles related to sustainability. He provided with the definition regarding the principles of sustainability being at a higher level along with an approach, which was considered to be generic. The principles that had been provided or proposed by him were:

- For the management of the various pertaining renewable sources, the rates of harvesting are required to possess egalitarian rates of regeneration (maintenance of the yield) as well as the rates of production of the wastes are required to be egalitarian with the natural volume or the capacities of the assimilative belonging to the ecosystems;
- The justifiable or the sustainable utilization of all the sources relating to the non-renewable resources generally tend to require an extensive investment in the process of having an apt utilization of such forms of the pertaining resources that were being coupled up with an investment which was considered to be equivalent in a substitute being renewable (for instance: the process of extraction of oil that has been paired up with the plantation of the trees for wood alcohol);
- Various pertaining technologies, which helps in raising the productivity of the various resources that means the quantity or the amount of value that has been taken out or mined in accordance with the per unit of resource) are required to get prioritized, rather than all those technologies which possess their main concentration on increment of the resource material (Daly, 1990).

The establishment of these sustainability principles at a higher level has shaped the way governments, industries, and communities approach sustainability.

This, being the principles is generally being introduced into the legislation of various pertaining countries, creating a point of beginning so that they could be able to provide with the definition to the strategies by the government or the various policies, which are being made for the general public.

2.2.3. Sustainable Manufacturing Practices

In accordance with many of the researchers, because of the nature belonging to the principles of sustainability, be it the theoretical or the abstract one the theoretical or abstract nature of the various practices have been established or incorporated.

Lately, an increment in the enumeration of the various institutions or the organizations which are integrating or connecting with the practices related to the sustainability into their respective functions or operations of manufacturing so that they could be able to have an improvement not only with their performance related to the environment but also with their respective performances belonging to the economy as well as society.

Therefore, all the practices that are being related with the SM generally tend to have been described in accordance with the opinion or the perspective being related with the environment.

Providing a reflection to this, Nordin et al., (2014) has seen the practices belonging to the SM as targeting towards the process of minimization of the various effects or the impacts of the operations related with the manufacturing based on the environment whilst the process of optimization of the efficacy or effectiveness of the production by the respective company.

It is worth mentioning that, before the term sustainability was coined, Bowen (1953) introduced the concept of corporate social responsibility (CSR).

For Bowen, the CSR has provided with a representation of the responsibility for various companies to take up the accountability for all the effect or the impact on the various domains of environment, be it ethical, social, human as well as ecological one, because of their respective practices, regarding the course of their specified activities within the pertaining businesses.

Hence, the CSR started getting acknowledged being the introduction as well as the execution or implementation of the growth or development of the sustainability under the tract of the process of management (Kechiche and Soparnot, 2012).

Similarly, Perez-Batrez et al., (2010, p. 193) provided with the statement that the norms or the concept regarding the CSR usually tends to have an inclusion of the terminology, i.e., sustainability.

While being similar, the CSR along with the sustainability can never be considered as same or identical. Both are considered to be the acts being voluntary corporate, which might possess distinct results or purposes. All the sustainability initiatives can be labeled as CSR initiatives, but some CSR initiatives cannot be directly related to sustainability.

The surge in the essentiality of the SM over the past so many years has caused an increment of the interest within the research or the study of practices being related with the sustainability.

Many researchers (for instance: Millar and Russell, 2011; Hurreeram et al., 2014; Habidin et al., 2015) possess quite sound understanding regarding the practices which are being associated with the SM within various distinct domains of industries along with the distinct countries.

Nevertheless, the vast majority of studies on sustainability practices have tackled mainly environmental practices, as well as the effects of sustainability practices on company performance, and sustainability practices among countries and sectors.

2.3. MODELING AND TACTICS FOR SUSTAINABLE MANUFACTURING: AN IMPROVEMENT METHODOLOGY

2.3.1. Research Methods

The various studies based on the research which has been conducted by various researchers is considered to part of an extensive project that helps in developing a tool related with the modeling along with the process of simulation.

It possesses a goal or target to present with an assistance to all the existing manufacturers so that they could be able to have an identification or the recognition of the various opportunities or chances related to the enhancements within their respective flow of Material, Energy along with the associated Waste resource having an apt utilization of the common or general tacts, a methodology for the enhancement along with the modeling of Material, Energy along with the associated Waste flows.

It tends to search out for the way of addressing the various questions based on such a research as to how general or common tacts are capable of providing assistance in the process of recognition or identification of the opportunities based on the pertinent enhancements being in a systematic way. This research was conducted in the two key facets or aspects:

- Building up of a theory having an apt utilization of the various strategies regarding SM along with the collection or the amalgamation of the various case studies from the literature or the self- collection; and
- The pertaining theory, which helps in assessing the partners belonging to the various industries with the help of THERM project.

Such cases that have been gathered as well as assessed or analyzed, were being segregated so that they could be able to have a sound understanding regarding the breadth of all such practices within the industry along with the knowledge of how all the other manufacturers are able to have an execution or implementation of likewise enhancements within their own respective factories.

All these practices were being evaluated or assessed under the lens of the model based on the varied norms or concepts of the pertaining ecosystem of manufacturing by having a great focus on the flows of Material, Energy along with the associated Waste sources being connected with the elements or the components of the three system (i.e., operations related to manufacturing, various facilities as well as buildings).

The generic tactics were then formulated to extract of mechanism of change and support the wide dissemination of these practices in the manufacturing industry.

A library based on the provided tact was certainly being made or created so as to make them accessible or feasible in accordance with the format, which possess the potentiality of getting instantly utilized with the help of the modeling tool that has been evolved or developed in THERM.

The process of gathering various practices has lately been expanded so that they could be able to broaden the range relating to the greatest of the practices being accessible within the database whereas the second facet possesses an inclusion of the applications based on the prototype regarding the model of manufacturing ecosystem.

The major involvement to the prevalence of the knowledge is certainly the formation or the creation of a library that has been wisely designed or structured related to the tacts which helps in the recognition or the identification of the great mechanism belonging to the enhancements and thus provides with the permit to the process of generalization of the practices belonging to the SM.

The contribution to practice is making tactics available to support manufacturers identifying improvement opportunities in a structured and systematic way.

2.3.2. Manufacturing System Modeling

The model of manufacturing ecosystem that are based on the concepts or norms set is grounded on the model of type II related to the Ecology of the various Industries: the input of the pertaining system (which means the entire intake or utilization of the various resources) along with its relevant output (the production or the emissions of pollutant as well as the waste materials, the resultant of the specific product that are being held within the technosphere) are generally being considered as restricted, along with the flow of the existing resources under the system hold a specified extent or degree of periodicity.

Thus, it provides with the meaning that the total of the entire pertaining flows under the system is usually being acknowledged as existing on higher level in comparison to the sum of the varied inputs along with their respective outputs to such forms of system, thus, having a decrement or reduction in the reliability or relativity of these systems on the various external resources as well as the sinks along with its impact or influence on the environment.

The provided model helps in presenting with three of the core elements or components belonging to the system of manufacturing having an inclusion of the following: providing assistance to the various facilities, operations related to the process of manufacturing, as well as the buildings, which are surrounding it.

All three components are linked by resource (material, energy, and waste) flows. Several strategies (or the set standards or norms or the principles), so as to have SM, were being gathered from certain conducted researches and thus can be described in brief in the following points:

- Ignorance or aversion of the utilization of various existing resource and thus, having an improvement in the efficacious or effectiveness of the process of transformation: utilization and wasting less by having a drastic or considerable increment in the productivity of several natural resources (be it the energy or the material);
- Closure of the loop of flow of the various pertaining resources: having a shift or a movement towards the pertinent models relating to the production that has been inspired or encouraged

by the biological factors, for instance: lessening up of the various resultants which are considered to be unwanted along with the transformation of such resultants into the inputs (having an inclusion of the waste energy): opting for the process of recycling along with all its alternatives;

- Automatization of the supply or replacing the pertaining technology: again investing in the certain natural capital with the help of replacing the input materials with the other ones: opting for non-toxic instead of toxic, picking up of renewable rather than the non-renewable resources;
- Shifting of the standards set: movement of these towards the various models of the business that are based on the solutions having an inclusion of all the designs or the structures related to the process of production as well as ownership that have been transformed: the varied systems based on the product service, structure of the chain of supply.

This model belonging to the ecosystem has often been utilized to provide with a description or definition to the required ways of all the occurring changes along with its respective aim or goal so that they could be able to shift towards having sustainability in the process of manufacturing.

Certain outlines have been drawn in respect to the gate of the factory. The work of the industries often tends to possess their core focus on the enhancement of the factory so as to have retention of the worth of all such pertaining resources along with the aversion or the ignorance of deterioration of the environment

The four strategies mentioned above are usually applied at supply-chain level beyond the control of a single company. This work takes a narrower view and applies the three first strategies at factory level.

The various components that have been modeled up are considered to be the buildings, the elements or components related to the technology (certain processes as well as the required equipment) that are being positioned in and are nearby the existent buildings, along with the process of connecting all the pertaining components belonging to the model with the help of flows of such resources, for instance:

- **Inputs:** All the energy as well as the required material having an inclusion of chemical and water;
- **Outputs:** The product along with the produced wastes consisting

of the process gathering all physical waste within the provided dust- bins and the waste of energy majorly in the form of heat.

2.3.3. Sustainable Manufacturing Tactics

The practices related to the SM were being gathered and were thus, assessed or analyzed so that they could be able to formulate all such common tacts.

The core objective or the target for any of the industry was to symbolize the pertaining principles or the procedures relating to such practices so as to make an application of these to the various distinct kinds or forms of technology as well as the resources. In turn, this supports the generalization of practices. SM practices were collected from two types of sources:

- Various researches that have been conducted in accordance with the principles along with the certain approaches required for the SM, which are, at times, based on the conducted survey related to the practices of various industries, or on the bases of assessment of all the recent activities practices.

Thus, such types of sources tend to present with a great range of the practices involved, but just a few information regarding them are being mentioned at the time of application of such practices or on the basis of the activities of the content based on the technicality.

- Various website on the Internet that is based on the best of practices, various instances from companies only can be provided. Such sources tend to present with much details regarding the activities along with the pertinent outcomes from the process of execution or implementation, but some of the details as to how all such enhancements were being able to get recognized or identified or what all the complexities were there that were being encountered at the time of implementing it.

These two forms of source have provided with the distinct knowledge or information regarding certain activities of the industries: many of the researches have presented with the entire reports based on the initial cost incurred on investment, cost incurred in the various operations along with the process of maintenance, as well as the yearly savings being in terminologies of energy, water, cost, and material whilst the various other cases provided with the information which was wholly inadequate or none of the required information was being provided regarding all the pertaining benefits of the execution.

Thus, it becomes quite complex for the researches to form or draw any of the conclusions that are supposedly based on the current patterns or trends in accordance with the scale of change, the expanse of hard- work that was being required or the scale of the savings.

Moreover, all collected cases reported success stories with no mention of challenges, difficulties or barriers to implementation, and no reported case of failure.

Ultimately, the third part provides with the recognition or identification of the certain well- defined responsibility so that they could be able to have an execution of all the enhancements occurring within the factory.

Likewise, to that of the second part, it is also being utilized in order to slender down the process of seeking up for the various practices towards specified well-defined arenas or tracts belonging to the company in consideration with the pertaining responsibilities of the respective people who are generally involved within the activities relating to the enhancements.

2.3.4. Improvement Methodology

The methodology, for having improvement in the industries, is required to follow a certain series that tend to connect or link up the various tacts to the utilization of all the data that has been processed in order to model the system of manufacturing. Interestingly, the order in which improvements can be identified does not follow the prioritization order presented earlier.

This, thus, presents with the core challenge regarding the process of growth or development of the various pertaining tools along with the methodology involved in the enhancement.

The main complexity involved in the process of recognition or identification of an improvement has not been reflecting the persistent hardship involved in the process of implementing it.

On the contrast, in many of the cases, the greater efforts are being required in the process of gathering or collection of the provided data so that they could be able to have an identification of the fruits that are hanging at a lower level (for instance: the equipment for stopping as well as repairing) whilst having a replacement of the certain components or elements belonging to the system at cost which is at higher level and that can be easily recognized identified instantly (for example: all the resources that have been black-listed or all the old equipment that have become ineffective).

Keeping this challenge in mind, this section presents the improvement opportunities following the prioritization order rather than the first possibility identified.

Thus, in order to have an access to the avoidance or aversion forms of enhancement, it becomes quite essential to notify various pertaining transformations within tacts that can be become real complex for anyone to have an identification as the researchers usually tend to possess the requirement if an expert knowledge regarding the process that is being involved in order to recognize various resources or processes that are being utilized quite gratuitously and thus, it can get eliminated.

The requirement for the improvements in having a reduction in the amount of waste generally tend to have its major focus on the resultant of such wastes so that they could be able to seek for a way out in order to have a reduction in pertinent losses or sustaining the worth of such resultant, even in case where it is being a waste usually includes the residuals, unwanted byproduct, and so on.

Such enhancements are basically being acknowledged as easy when being compared as they are being permitted to have a rapid savings within the various resources as well as the cost incurred with the restricted labors.

However, manufacturers' knowledge about their waste is often limited, and for the waste patterns to be identified, a thorough data collection must be conducted. The major concentration will certainly be on all those processes, which are being considered as the consumers of an extensive resource along with the generators of waste materials.

The utilization of a tool related to simulation has been acknowledged as an essential asset to provide with permit to the well- defined search for the waste that is compatible as well as the requirement within the system having a consideration of the criticality of the systems that has been modeled up, the timing belonging to the flows along with the longitudinal dimension.

They involve large-scale changes by improving the source of supply and using high-efficiency technology, but they also reduce more dramatically the environmental impact of the manufacturing activities.

All the tacts that have been associated or connected with the required database relating to all the best practices so that they could be able to provide with a suggestion regarding the substitute resources or solutions based on certain technologies.

2.4. SUSTAINABLE MANUFACTURING TACTICS AND CROSS-FUNCTIONAL FACTORY MODELING

Industry has conventionally been linked or connected with the pessimist effect or influence on the environment: since the last few years, the deterioration of natural environment because of the growth or rise of the population and the increment that has been linked within the utilization of such resources (Holdren and Ehrlich, 1974), the growth or development of the economy along with the process of connected amplification of all the activities that are involved within the various industries (Meadows and Club of Rome, 1974) have become an extensive problem or issue of the globe that cannot be denied (World Commission on Environment and Development, 1987).

With the need for sustainability now widely recognized as a great challenge for society, industrial companies have become part of the solution to change the way society operates (Erkman, 1997; Jovane et al., 2008).

2.4.1. Research Program

The work that has been presented within various researches that have been conducted is considered to be part of a broader project which is often known with the help of life energy along with the process of modeling up of resources (THERM Project, 2011) that possesses an objective of providing an assistance to the enhancement or improvement to the SM (Ball et al., 2011).

The work is being considered as quite pragmatic as well as influencing. It generally tend to begin with the evolvement or development of a specified library of tacts that has been organized and designed or structured in accordance with an enhancement of the provided hierarchy that has been emerged from the pertaining hierarchies belonging to the waste or the energy as well as sets of standards or principles based on the sustainability of the process of manufacturing, various concepts along with the strategies.

The tactics aims at bridging the gap between high-level concepts and observed industrial practices for SM.

It has been acknowledged as an approach that is based on the certain steps being in sequence, grounded on the modeling up of factory that helps in the process of incorporation of the library that has already been designed belonging to the certain tactics so that they could be able to have an improvement within the flow of the resources by witnessing the factory in terms of an ecosystem. A factory modeling prototype tool integrating buildings, facilities, and manufacturing operations is presented to test the integrated methodology.

2.4.2. Improvement Hierarchy and Tactics for Sustainable Manufacturing

Under this section, various findings of the research have been provided in the format of a library related to certain tacts in order to have SM.

The library related to the tacts has been designed or structured by making a great utilization of the hierarchy of enhancements that generally tends to prioritize the various available alternatives. SM tactics were formulated based on the mechanisms of change observed in practices collected and analyzed in a previous study (Despeisse et al., 2012b).

There pertain various formulations belonging to the category of verb-noun so as to provide with the specification regarding the kind of transformation, viz: replacement, optimization, removes, adds, and so on, and thus, the main focus related to the change viz: the flow of the resources along with the technology.

In here, it has been said that tactics are being considered as either common or general enough so as to get applied in the various pertaining environments as well as the disciplines, which tends to lead to the various specified enhancements within certain levels of processes.

2.4.3. Need for Quantitative Analysis

Requirement for the analysis in quantitative terms exists as they are required to have an assessment or evaluation of the various influences or the effects of the activities related to the process of manufacturing on the environment along with the pros or advantages of all the probable enhancements. The analysis can be applied at different resolution levels to derive opportunities incrementally as effort is increased.

The already existent tools related to the modeling generally tend to present with an analysis of energy in the process of building up of modeling (Clarke, 2001; PérezLombard et al., 2009), flows of the respective product along with the timing of flows of the pertaining process in manufacturing (Pandya, 1995), but none amongst them tend to cover up the entire facets or the phases to have an account for all the specified flows of the resource, recurrence of certain processes along with the spatial dimensions.

In accordance with the approaches being lean or green as well as the tools relating the modeling of the process of manufacturing, various novel techniques along with the methodology tends to possess a requirement of the growing developments to get new methodologies and techniques require incremental development to be advanced and thus, in order to have an inclusion of whole of the provided components that are being required so as to provide with assistance to the pertaining structures or design and assessment of the various systems related to the SM (Jahangirian et al., 2010).

The analysis has to account for location and time in a manner that is not supported by either manufacturing process simulation tools or building energy tools.

2.5. MODELING AND ANALYSIS OF SUSTAINABLE MANUFACTURING SYSTEM USING A DIGRAPH-BASED APPROACH

Many of the researchers have provided with the various statements. They were entirely based on the sound understanding that, they have gotten from the countries that have already been developed. Further, for developing countries like India, foreign investment and globalization forces manufacturers to manufacture products for customers in developed countries (Gunasekaran and Spalanzani, 2012).

The growth that has been persistent in case of stability can be provided with an assurance by having evaluation as well as the interaction of the modeling amongst the four provided dimensions. The evaluation of the stable growth is being considered as quite a complex norm and thus, has been linked or connected with the ambiguity.

2.5.1. Global Sustainable Institutions

All across the globe, there exists many of the organizations or the institutions which all possess various contributions that were being considered as quite extensive in the direction of growth or development of the set standards or norms, certain methodologies along with the processes that are required for the evaluation or assessment, examination as well as the various facets belonging to the process of reporting of the sustainability.

Apart from these, the standardization for assessment of process and product impact assessment were outlined in Life Cycle Assessment (LCA)

(ISO 14040, 2006 and ISO 14044, 2006). Thus, if one wants to have a measurement, as well as the monitoring of the required progress in the direction of the production, being sustainable, many of the researchers have provided with certain pertaining set of indicators or signals in order to provide with a support or aid to the various organizations having great manufacturing in the direction of the process of setting up a gold standard for the performance of the environment.

LCA provides with an aid so that they could be able to demonstrate the performance of the environment with the help of a compendious set of categories having a great effect or impact on the environment. The impact categories vary according to the characterization model. The detailed LCA methodology was outlined in ISO 14040 (2006) and 14044 (2006).

2.5.2. Digraph Modeling for Criteria under Economic Sustainability Dimension

Below a digraph has been provided that has been developed in accordance with the criteria being present under dimension of the sustainability within the respective economy. Descriptions regarding the interaction amongst such a criteria and explanations for interactions among criteria have been provided below:

- **Market Potential (MP):** The criterion related to the Flexibility along with the Revert to the Changes or transformations (FRC) MP tends to provide with a scope of appreciation related to the organization in a specified market (Lei, Hitt and Goldhar, 1996).

Where, the Flexibility, Revert to the certain Changes often deals with the potentiality possessed by any of the pertaining organizations to provide with a response to the market, which is being considered as dynamic, volatile as well as uncertain.

- **Streamlined flow (SF):** The criterion related to the Defect Ratio (DR) SF usually possess its main focus on forming up of the flows that are swift and pertains thru the organizations or the various institutions without the possession of any kind interruptions (Sullivan, McDonald and Van Aken, 2002).

SF tends to have a removal of the varied wastes that are being connected or linked with the operations related to the process of manufacturing (Figure 2.3).

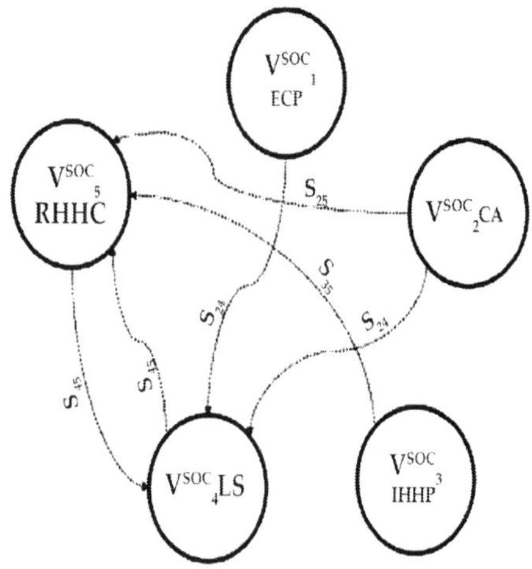

Figure 2.3: Digraph pertaining to social sustainability.

Source: https://www.researchgate.net/publication/322130994_Modelling_and_analysis_of_sustainable_manufacturing_system_using_a_digraph-based_approach

2.6. SUSTAINABLE MANUFACTURING: EVALUATION AND MODELING OF ENVIRONMENTAL IMPACTS IN ADDITIVE MANUFACTURING

Sustainability holds an essential position in the world of industries similar to that of the pertaining societies.

And, this is the reason as to why the researchers have taken into account that at the time of the steps related to the manufacturing, certain facets related to environment have provided with a permit to the manufacturers so that they could be able to differentiate from all their respective competitors so as to have an increment in their profitability along with the competitiveness.

This method of quantification has been normalized by two agencies: SETAC (Society of Environmental Toxicology and Chemistry) and UNEP (United Nations Environment Programme) under the standard ISO 14044 (Figure 2.4).

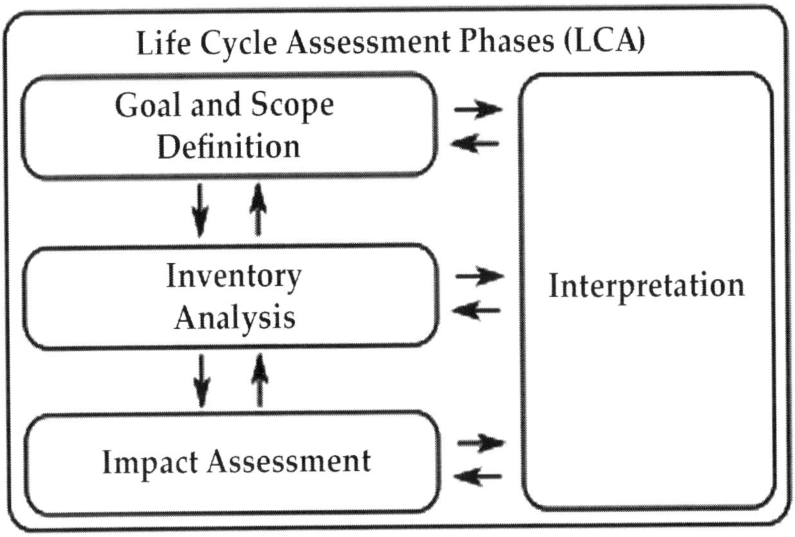

Figure 2.4: Life cycle assessment phases.

Source: https://hal.archives-ouvertes.fr/hal-00881866/file/HAL_Sustainable_manufacturing_Evaluation_and_Modelling_of_environmental_impacts_in_additive_manufacturing.pdf

2.6.1. State of the Art

Environmental considerations are taking an important place in the manufacturing world. In fact, green manufacturing is defined as the first step towards sustainable development.

In consideration of the field related to the process of machining along with the removal of the chip, many of the works have been done in accordance with the consideration of pertaining environment.

Many of the researchers have provided with approach that are analytic in nature so that they could be able to have a strong determination of the various effects or influences on the environment at the time of cutting of the technologies.

Wherein, Avram, et al., has presented with a methodology that was a pre-process and which tends to facilitate the determination of the consumption or the utilization of the electricity of the process of turning the provided machines.

Similarly, Kong et al. formed up many of the software tools, which provided with the assistance in having an approximate measure of the utilization of the energy of the tool related to the machine.

2.6.2. Global Analysis

To provide with a novel methodology or procedure in order to have an assessment of the effect or influence on the environment, many of the researchers have segregated their research into three distinct steps:

- **Step 1**: Description regarding the limits that has been put in the research;
- **Step 2**: Assessment of the utilization of all the flows; and
- **Step 3**: Transformation of the utilization of the flows into the influence or effect on the environment.

In a first time, as it is mentioned previously, the limits of the study will be clearly defined; otherwise, it is necessary to consider all products used in part manufacture, such as:

- Manufacturing of powder;
- Creation of the Inert gas;
- Production of the hydraulic fluid;
- Production of Compressed air;
- Creation of the electricity.

In this paper, the study will be focused on the consumption during the process before to be extended it later to the powder production, part recycling, etc.

This research has been obtained on a Distributed Linear Algebra Machine (DLAM) that is being composed with a great set of standards or principles which tend to follow the pertaining characteristics:

- A fiber laser having a 4 kW;
- Two pertaining feeders of the powder;
- Three axes that are linear;
- Two axes which can be rotated.

Secondly, the process belonging to the utilization of all the flow has been assessed. The evaluation of the provided input relating to the machine generally tends to provide with a permit so that they could be able to divide the utilization of the pertinent flows that are distinct into three provided parts:

- Consumption of electricity;
- Apt utilization of the provided material; and
- Gobbling up of the certain fluids.

The provided three sorts tend to provide with a permit to have a characterization of the already settled parts of the processes but also helps in encouraging them in reckoning the influence or the effect on the environment across the globe presenting with the rating in the third sort or part. In this third step of global E.I. assessment, two computations are realized.

2.7. SUSTAINABLE MANUFACTURING AND DESIGN: CONCEPTS, PRACTICES, AND NEEDS

Sustainability has been considered as a requirement or the need, which is becoming progressively essential for the various activities of the human beings, which thus, makes the growth or development of sustainability a core goal in the evolvement of the human beings (Figure 2.5).

Figure 2.5: Sustainable manufacturing (SM).

Source: https://www.mdpi.com/2071-1050/4/2/154

2.7.1. Background: Sustainability and Sustainability Indicators

In the process of providing a description, grasping various informations and thus, implementing the SM, it is generally being considered as quite essential to possess a great knowledge regarding sustainability along with the various pertaining signals or the indicators for this purpose. In general, sustainability is an imprint that got its description in the provided manner and thus, has a hold of distinct meanings in consideration with the distinct people.

The process of having sustainability in manufacturing can be seen as owning three major parts, which are based on the society, economy along with the environment, thus, having an inclusion of the political aspect as well (Figure 2.6).

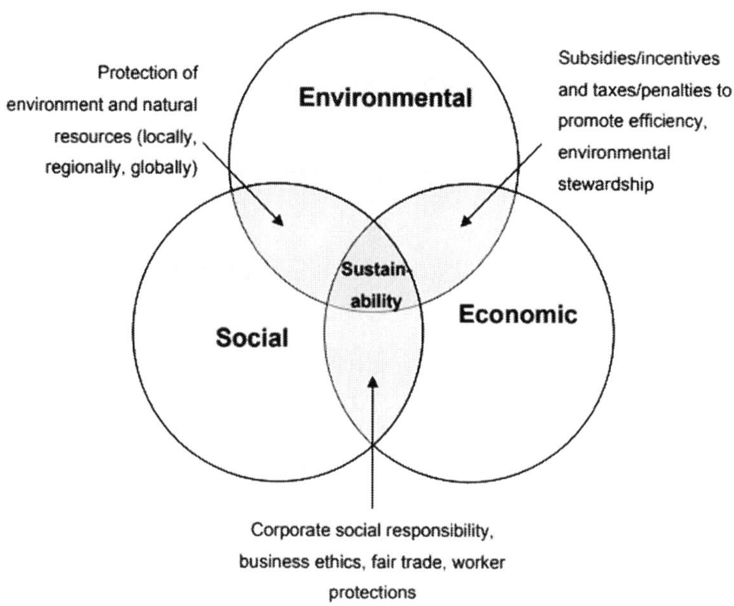

Figure 2.6: Sustainability as the intersection of its three key parts, and examples of features at the intersection of any two parts.

Source: https://pdfs.semanticscholar.org/5cbb/248d2f6b6426baee81a7d1037cd1151e3400.pdf

2.7.2. Sustainable Manufacturing (SM)

Having an improvement within the sustainability as well as stewardship whilst possessing a maintenance of the productivity along with the profitability, have been seen as the goals related to the strategy belonging to the companies having SM.

2.7.2.1. Manufacturing and the Environment

Conventionally the strategies that were being made for the process of manufacturing might have taken into consideration the comparisons relating to the process of production for a great variety of the matrix belonging to varied products.

Population is being considered as quite complex to have pressurize as well as having prosperity that has been highly required by the general people.

2.7.2.2. Manufacturing and Sustainability

The development or the evolvement of the SM from the varied conceptions belonging to the growth or development of the sustainability that has been formed in the year 1980 so that they could be able to refer the various concerns or worried related to the influence or the effect on the environment, evolvement of the economy, the process of globalization, pertaining discrimination and many other related factors.

The production possessing sustainability got first introduced in the year 1992 at the conference of UNCED in Rio de Janeiro being a form of a guide in order to provide with the aid to the existing companies along with the transitions within the governments in the direction of the development of the sustainability in the process of manufacturing.

Conduction of the research within the pertaining tracts or the arenas has been constant by so many.

The method that has been described in the latter part by the various researchers has been acknowledged as the case of the business for the having sustainability in the process of manufacturing that tends to have an involvement of varied holistic approaches that tends to depend on the various standards as well as the principles that have been set related to the possession of prosperity within economy, stewardship of the respective environment as well as taking up of the responsibility in the corporate world.

REFERENCES

1. Alayón, C., (2018). *Exploring Sustainable Manufacturing Principles and Practices* [eBook] (p. 98). Retrieved from: https://www.diva-portal.org/smash/get/diva2:1038430/FULLTEXT01.pdf (Accessed on 29 March 2019).
2. Ball, P., & Smith, L., (2012). *Steps Towards Sustainable Manufacturing Through Modeling Material, Energy and Waste Flows*. Retrieved from: https://www.sciencedirect.com/science/article/pii/S0925527312000515 (Accessed on 29 March 2019).
3. Carley, D., (2014). *Success Paths to Sustainable Manufacturing.* [eBook] (p. 33). Retrieved from: https://spea.indiana.edu/doc/research/sustainability–2014.pdf (Accessed on 29 March 2019).
4. Despeisse, M., Ball, P., & Evans, S., (2018). *Modeling and Tactics for Sustainable Manufacturing: An Improvement Methodology*. Retrieved from: https://www.researchgate.net/publication/278710946_Modelling_and_Tactics_for_Sustainable_Manufacturing_An_Improvement_Methodology (Accessed on 29 March 2019).
5. Despeisse, M., Ball, P., & Evans, S., (2018). *Sustainable Manufacturing Tactics and Modeling: An Improvement Methodology for Manufacturers*. Retrieved from: https://www.researchgate.net/publication/323253652_Sustainable_Manufacturing_Tactics_and_Modelling_An_Improvement_Methodology_for_Manufacturers (Accessed on 29 March 2019).
6. Despeisse, M., Oates, R. M., & Ball, P. D., (2018). *Sustainable Manufacturing Tactics and Cross-Functional Factory Modeling* [eBook] (p. 21). Retrieved from: https://dspace.lib.cranfield.ac.uk/bitstream/handle/1826/8019/Sustainable_manufacturing_tactics–2013.pdf?sequence=1&isAllowed=y (Accessed on 29 March 2019).
7. Jain, S., Shao, G., Brodsky, A., & Riddick, F., (2013). *A Model-Based Continuous Improvement Methodology for Sustainable Manufacturing* [eBook] (p. 10). Retrieved from: https://pdfs.semanticscholar.org/634a/6aadb443f6c90fa60c4b6fc3d0ea0fb9fae5.pdf (Accessed on 29 March 2019).
8. Kibira, D., & McLean, C., (2018). *Modeling and Simulation for Sustainable Manufacturing* [eBook] (p. 10). Retrieved from: http://citeseerx.ist.psu.edu/viewdoc/download?doi=10.1.1.161.2258&rep=r

ep1&type=pdf (Accessed on 29 March 2019).

9. Le Bourhis, F., Kerbrat, O., Hascoet, J., & Mognol, P., (2018). *Sustainable Manufacturing: Evaluation and Modeling of Environmental Impacts in Additive Manufacturing* [eBook] (p. 22). Retrieved from: https://hal.archives-ouvertes.fr/hal–00881866/file/HAL_Sustainable_manufacturing_Evaluation_and_Modelling_of_environmental_impacts_in_additive_manufacturing.pdf (Accessed on 29 March 2019).
10. Ocampo, L., & Clark, E., (2018). *A Sustainable Manufacturing Strategy Framework: The Convergence of Two Fields* [eBook] (p. 29). Retrieved from: http://web.usm.my/aamj/20022015/aamj20022015_2.pdf (Accessed on 29 March 2019).
11. Rosen, M., & Kishawy, H. A., (2012). *Sustainable Manufacturing and Design: Concepts, Practices, and Needs*. [eBook] (p. 21). Retrieved from: https://pdfs.semanticscholar.org/5cbb/248d2f6b6426baee81a7d1037cd1151e3400.pdf (Accessed on 29 March 2019).
12. *Sustainable Manufacturing – the Next Steps,* (2018). [eBook] (p. 24). Retrieved from: http://file:///C:/Users/ADMIN/Downloads/SSD2388%20Sustainable%20Manufacturing%20Insight%20Report-v5%20LR.pdf (Accessed on 29 March 2019).
13. *Sustainable Manufacturing Unpacking the Issues*, (2009). [eBook] (p. 32). Retrieved from: http://www.mskills.com.au/DownloadManager/Downloads/Sustainable%20Manufacturing%20-%20unpacking%20the%20issues.pdf (Accessed on 29 March 2019).
14. Taghavi, N., (2015). *Sustainable Manufacturing Strategy; Identifying Gaps in Theory and Practice*. [eBook] (p. 70). Retrieved from: http://publications.lib.chalmers.se/records/fulltext/223923/223923.pdf (Accessed on 29 March 2019).
15. Vimal, K., Vinodh, S., & Gurumurthy, A., (2018). *Modeling and Analysis of Sustainable Manufacturing System Using a Digraph-Based Approach*. Retrieved from: https://www.researchgate.net/publication/322130994_Modelling_and_analysis_of_sustainable_manufacturing_system_using_a_digraph-based_approach (Accessed on 29 March 2019).

CHAPTER 3

NANOTECHNOLOGY FOR SUSTAINABLE MANUFACTURING

CONTENTS

Nanotechnology has the power of playing an essential role in enhancing the sustainability of a broad variety of industrial sectors. Nanomaterials can majorly contribute to more sustainable manufacturing (SM) by cleaner, quite less inefficient processes of production and also can substitute traditional materials that lead to saving in raw materials as well as energy. Nanotechnology for SM includes recent development in the fields of energy and materials proficiency concerned with the resource saving along with the raw materials conservation, both of which are the main drivers for the use of nanotechnology in the industrial areas. The chapter describes the application of nanotechnology in various sectors such as electronics, food processing, aerospace, medicine, etc. it also discusses the opportunities as well as challenges related to the use of nanotechnology in SM.

3.1. INTRODUCTION

In the past few years, academia, industry as well as government have practiced an enhancement in various research activities along with the expenditures in the area of nanotechnology. For instance, the expenditures of the US government for the research and development purpose related to nanotechnology have increased by 35.6% from the year of 2000 to 2004. Whereas the overall investment has increased by 87.4% at the international level, led by Japan as well as Western Europe (Rocco, 2003).

The developing field of nanotechnology is likely to lead to extraordinary considerate as well as control of the matter composition (Tsakalakos et al., 2003). It has already been stated that eventually, nanotechnology will transform the society in numerous ways mainly in the design as well manufacturing levels of the development of product (NSTC, 1999).

The basis of motivation for discovering nanotechnologies is the unusual physical, chemical as well as biological characteristics which matter exhibits at the nanoscale (10^{-9} meter). Few of the advantages are strength, enhanced thermal conductivity, hardness, and higher rates of catalytic conversion (Theodore and Kunz, 2005).

From the societal outlook, these kinds of characteristics could be demoralized to enable the success of the United Nations Millennium Development Goals (Salamanca-Buentello et al., 2005) by the development of different sustainable technologies (EPA, 2007). In the year 2003, it was assumed that by the year 2006, almost half of the new products would comprise few types of nanotechnologies.

It is yet very difficult to prove such kinds of statements. The capacity of nanotechnology is such that around 2 million workers are likely to be required in the industry of nanotechnology in upcoming years, along with the revenues of one trillion dollars around the world.

In consideration with the industry of nanotechnology, it can be compared with the industry of automobiles that had global revenues of one trillion dollars in the year of 1998 (Kaku, 1998). These kinds of promises have attained the attention of various regions or countries around the globe comprising Europe, Asia, and the United States.

The European Union (EU) research commissioner specified that the field of nanotechnology is the key of attaining sustainable development (SD) and linked with completing the needs of the Kyoto Protocol (IOP, 2002).

In spite of the support of government, very little is known regarding how to develop advanced nanotechnologies in such a way that assures the notion of SD. In the period of environmental movement because of the social awareness as well as increased demand of clean technologies, nanotechnologies must be observed at through c complicated lens for the aim of taking the best it can give while neglecting the potential negative consequences.

The development, as well as establishment of nanotechnology, coincides with the period of resource optimization, issues of manufacturing wastes, high cost of energy supply and the concerns related to emissions in the atmosphere that have been linked with the negative impacts on the biodiversity, environment, and human health.

These situations give rise to a major question that, 'should the field of nanotechnology be followed in the absence of a developed knowledge base required to regulate such advanced technologies?' Various manufacturers face so many issues and challenges related to sustainability that varies from enhancing cost of energy to the management of waste.

Therefore, different approaches have been taken into consideration in the past ten years so as to reduce the negative environmental influences that confront numerous manufacturers. Few of the issues and challenges comprise: providing workers or employees with a safe and secure environment of working as well as outsourcing of manufacturing to the industrializing nations which has enhanced to record levels in various nations over the last ten years.

It is the fact that the field of manufacturing has the second highest number of populations employed and therefore, it has a huge impact on the society. These facts illuminate the link between the society, industry, and environment.

Therefore, it validates the status of addressing the manufacturing sustainability. The main objective of this chapter is to discuss regarding the applications as well as potential of nanotechnology in various fields of manufacturing to contribute to the desire to attain SD at global level specifically to the industry of manufacturing.

Every individual requires food, energy, water, clothing, shelter, healthcare, etc., to survive as well as prosper on the Earth. One of the major problems facing society in the 21st century is providing standard living conditions to every human being while reducing the effect of human activities on the climate as well as environment, as the population of the world reaches around 8–10 billion by the year 2050 (Diallo et al., 2013).

According to the Brundtland Commission of the United Nations, SD can be defined as "that which meets the needs of the present without compromising the ability of future generations to meet their own needs" (Brundtland, 1987).

These days, the world is facing major issues and challenges to meet up the increased demands of the basic commodities such as water, food, and energy, finished goods such as cell phones, cars, airplanes, and services such as healthcare, shelter, and employment while decreasing the greenhouse gas emissions along with the environmental footprint of industry and agriculture (Godfray et al., 2010; Diallo and Brinker, 2011; Brinker and Ginger, 2011; Diallo et al., 2013).

Just after the beginning of the NNI that is National Nanotechnology Initiative, it was proposed that the field of nanotechnology can provide various sustainable solutions to the universal issues and challenges facing society.

Roco, during a presentation at the Cornell Nanofabrication Center, discussed that how nanotechnology could help to enhance yields of agriculture, provide much efficient as well as cost-effective treatment of water and desalination technologies, and aid the development of clean as well as renewable sources of energy comprising highly effective solar photovoltaic cells.

As the field of nanotechnology remains to develop, the agenda of research is mainly concentrated on addressing majorly two questions concerned with the sustainability over the next ten or twenty years:

- Can nanotechnology helps to address the issues and challenges of enhancing the global sustainability in food, water, energy, shelter, healthcare, transportation, and employment?
- Can nanotechnology be established in a sustainable way with huge societal advantages and little impact on the global climate and environment of the Earth?

3.2. NANOTECHNOLOGY FOR SUSTAINABLE MANUFACTURING (SM): SOCIETAL PERSPECTIVES

Brundtland's Commission called the three pillars of sustainability social, economic, and environment, suggesting that sustainability requires convergence of all three (Brundtland, 1987). Therefore, two significant aims of global sustainability are to attain inclusive economic as well as societal development that benefits every individual while reducing the impact of activities done by humans on the climate and ecosystem of the Earth (UN, 2012; Diallo et al., 2013).

There are various authors who make the case that nanotechnology-based solutions are generally projected without any serious consideration of other alternatives or few negative side effects of these kinds of technologies.

In 2012, Wiek et al., combined the outcomes of various discussions form literature reviews and interdisciplinary workshops to assess the utilization of nanotechnology to address three urban sustainability issues in the Gateway Corridor Community of Phoenix in Arizona (USA):

- Contamination of water;
- Utilization of energy; and
- Childhood obesity.

Various authors reported that nanotechnology has the potential to make major contribution for a sustainable future, but those kinds of interventions should be tied as well as embedded in a systemic intervention strategy that are not wholly dependent on nanotechnology as the silver bullet.

As sustainability involves considering social, economic, and environmental aspects, it is very crucial in every case to combine fundamental science such as synthesis of material, characterization, modeling along with engineering research such as designing of system, fabrication, testing, commercialization, and few societal benefits such as new employment and cleaner environment (Diallo and Brinker, 2011; Diallo et al., 2013).

Therefore, the solutions of nanotechnology for the SD cannot be simply addressed at the stage of small as well as single investigator funded research grants. Sustainability research and development has to be combined with bigger research aims and comprised from the starting in huge interdisciplinary programs to be carried out by various interdisciplinary teams of investigators as well as focused government-funded centers of research and development.

Various researchers have reviewed the policies of nanotechnology for SM in the past ten years. They reported that various key policy enactments encouraging fundamental, as well as application-driven, aided various countries to lead in context of nanotechnology research and development along with the USA, Germany, and Japan.

The universal sustainability issues facing the world are complicated and hence comprise various interdependent areas. Nanotechnology has evolved as a multipurpose platform for the establishment of technical solutions to universal sustainability issues facing the world. The major advances of nanotechnology under the areas of global sustainability are:

- Purification of water;
- Technologies related to clean energy;
- Management of greenhouse gases;
- Materials supply as well as utilization;
- Green manufacturing, chemistry; and
- Societal perspectives.

Numerous retrospectives approve that nanotechnology continuously provide unprecedented chances to make functional materials for the sustainable technologies or products with higher optical, electronic, magnetic, and catalytic properties.

The ideal kind of nanomaterial can be processed into different form factors consisting of water-soluble supramolecular hosts, fibers, particles, and membranes. Therefore, they have evolved out as building blocks for a wide range of applications of sustainability such as:

- Treatment of water, its reuse, and desalination;
- Generation of energy, its conversion, and storage;
- Capturing of carbon dioxide, its storage, and conversion;
- Environmental monitoring and remediation;
- Material extraction, purification, and recovery; and
- Manufacturing.

Sustainability can be determined by the coupled interactions between the growth of population and human requirements, cultural, and societal values, and the human-built boundaries of earth's system and environment.

Thus, it is vigorous to consider the complicated linkage between the social system that is the institutions that provide support to the existence of human on Earth, the global system that is the ecosystem of earth that support life of humans and the human system that is all of the other factors that affect the health as well as well-being of humans, to attain global sustainability.

After the Rio Conference of June 2012 on SD, the United Nations started developing its developing agenda post-2015. Three major guiding principles of the new agenda are equality, human rights, and sustainability.

The execution of the United Nations development agenda will need few transformative changes in the methods goods are produced and consumed, natural resources are managed, and our society is governed. Additionally, the convergence of knowledge, technology, and society will also be needed to produce the revolutionary technology and transformative advances crucial to realizing a healthy, secure, sustainable, and peaceful world.

It is expected that, nanotechnology will remain as the most crucial component of the convergence of knowledge, technology, and society as it provides the potential to increase the limits of sustainability and to address all of the crucial requirements of human development on the Earth comprising basic commodities such as food, water, and energy, finished goods such as cars, cell phones, and airplanes and various services such as healthcare, shelter, and employment.

Although, it is crucial to be assure that any potential side effects of nanotechnology on the health of human as well on environment are significantly assessed and addressed prior to the large-scale deployment of nanotechnology-based products and solutions for the global sustainability issues and challenges facing the world in coming next twenty years.

3.3. NANOTECHNOLOGY AND THE FUTURE OF ELECTRONICS

Nanotechnology is significantly changing the area of electronics mainly in regard to optics, computers, and telecommunications. This section deals with the applications of nanotechnology in the field of electronics.

Nanotechnology deals with all the system that are designed as well as manufactured at the scale of the atom or the nanometer scale. Especially, nanotechnology comprises the manipulation as well as control of atom and molecules that are building blocks of all materials.

Broadly, there are two major approaches in nanotechnology that are top-down and bottom-up. The first approach, top-down approach infers the process of controlling to force atoms and molecules to build up themselves to wanted structures or locations. The second approach, the bottom-up comprises manipulating few numbers individual atoms or complicated molecules, into structures naturally utilizing minute probes.

Nanotechnology is being used by the industry of electronics, and it is surprising to know that various of electronics have already combined various applications that the nanotechnology science has established. For instance, advanced computer microprocessors have less than 100 nm features. Small size means a substantial increase in the speed and much processing ability.

Undoubtedly, these kinds of advances will aid to attain better computers. However, in the coming future, the current technology of electronics will no longer be sufficient to handle the demand of advanced chips microprocessors.

Etching or lithography is also known as the method of chip manufacturing. By using this technology, a probe actually writes over the surface of the chip circuit. This method of circuit building in electronic chips has a restriction of about 22nm. The advanced chip processors utilize 60–70 nm size features.

Below 22 nm, errors will take place, short circuits and silicon restrictions will avoid manufacturing of chip, and nanotechnology may provide advanced methods of working for electronics. Nanotechnology science is making new materials of circuit, advanced processors, advanced ways of storing information as well as advanced manners of information transfer.

Nanotechnology is utilized in various communications, electronic, and computing applications. It provides smaller, faster, and more probable systems. These systems store as well as manage huge amount of information.

Nanoelectronics define the use of nanotechnology in electronic components. There are various applications like electronic and computing products comprise the keyboard, antimicrobial, and antibacterial coatings on the mouse, the castings of cell phone and Flash memory chips for iPod (Figure 3.1).

Figure 3.1: Design of future nano-electronic circuit.

Source: https://genesisnanotech.wordpress.com/2017/08/22/the-design-of-future-nano-electronic-circuits-free-flowing-electrons-in-graphene/

The main objective of Nanoelectronics is to process, transmit, and store information by taking benefits of the matter's properties that are very different from that of the macroscopic properties. Nanotechnology is basically used for printed electronics, smart cards, and smart packaging. It is more life-like video games as well as flexible displays for e-book readers.

Nanotechnology is majorly used for the nanoscale transistors that are more powerful and faster and mainly the energy efficient. In the coming future, the entire memory of computer will be stored on a single small chip. It is being used in advance televisions, computers, laptops, cell phones, digital cameras and also in various devices having nanostructured polymer films that are called an organic light emitting diodes or OLEDs. The OLED screens provide longer lifetimes as well as brighter consumption. Nanotechnology is usually utilized for MRAM (Magnetic Random-Access Memory) allowed by the nanometer scale magnetic tunnel junctions that can effectively as well as quickly save encrypted data during the period of crash or system shutdown. Hence, it enables the features of resume play.

In electronics, nanotechnology provides smaller, faster, and improved hand-held devices. It also provides modern display technologies with data storage, conductive nanomaterials, printable, and flexible electronics, magnetic nanoparticles for data storage and quantum computing.

Nanotechnology actually can revolutionize various products of electronic along with the procedures as well as applications like electronic products comprise Nanotransistors, OLED, Nanodiodes, quantum computers, plasma displays, etc.

In context with the electronics, nanotechnology enhances the abilities of electronic devices while decreasing their weight as well as power consumption. It also enhances the density of memory chips and decreases the size of transistors which are used in ICs that is integrated circuits.

3.4. NANOTECHNOLOGY IN FOOD PROCESSING

The food ingredients that are of nanoscale are being developed with the rights that they provide enhanced taste, consistency, and texture. Nanotechnology increases the shelf life of various food items, and it also help to brought down the level of food wastage because of the microbial infestation (Pradhan et al., 2015) (Figure 3.2).

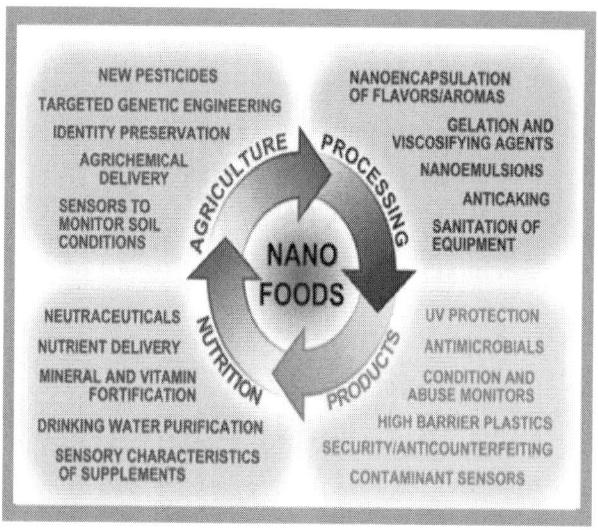

Figure 3.2: Nanotechnology in processing of food.

Source: https://www.slideshare.net/YAMUNAKURIAN/nanotechnology-in-food-processing-and-food-packaging

These days, nanocarriers are majorly being used as the system of delivery to carry additives of food in all of the food products without altering their basic morphology. The size of particle may affect the delivery of bioactive compounds to different sites in the body as earlier it was noticed that in few cell lines, just submicron nanoparticles can be efficiently absorbed but not the bigger size microparticles (Ezhilarasi et al., 2013).

An ideal system of delivery is supposed to have the following properties:

- Capable of delivering the active compounds exactly at the place of target;
- Assure availability at a specific rate and at a target time; and
- Effective to sustain active compounds at appropriate levels for extended period of time.

Nanotechnology being utilized in the formation of emulsions, encapsulation, simple solutions, biopolymer matrices, and association colloids provides effective systems of delivery with all of the above-defined characteristics.

Nanopolymers are now trying to substitute traditional materials in food packaging. Nanosensors can be utilized to verify the presence of mycotoxins, microorganisms, and various other contaminants in food (Bratovčić, 2015).

Nanoparticles have much better properties for the process of encapsulation and release efficiency as compared to conventional systems of encapsulation. Nanoencapsulation mask tastes or odors, control the release of the active agents, control interactions of the active ingredients with the matrix of food, assure availability at a specific rate and at a target time, and secure them from heat, moisture, chemical or biological deterioration during processing, storage, and utilization and display compatibility with some other compounds in the system.

Additionally, these systems of delivery have the ability to deeply penetrate into tissues because of their small size and hence, permit efficient delivery of active compounds to the specific target sites in the body (Lamprecht et al., 2004). Different kinds of natural and synthetic polymer-based encapsulation systems of delivery have been expanded for the enhanced bioavailability as well as preservation of the active food components.

3.4.1. Nanotechnology in Food Packaging

A required material of packaging should have gas as well as moisture permeability linked with the strength and biodegradability (Couch et al., 2016). Nano-based active and small food packaging confer numerous benefits over traditional methods of packaging from providing best material of packaging with enhanced barrier properties, mechanical strength, antimicrobial films to nanosensing for detection of pathogen and alerting all of the consumers to the safety status of food (Mihindukulasuriya and Lim, 2014).

Use of nanocomposites as an active material for packaging as well as material coating can also be utilized to enhance the packaging of food (Pinto et al., 2013). Various researchers were involved in studying the antimicrobial characteristics of organic compounds such as organic acids, essential oils, and bacteriocins, and their utilization as antimicrobial packaging in polymeric matrices (Gálvez et al., 2007; Schirmer et al., 2009).

Though, these kinds of compounds do not properly fit into the various steps of food processing that require high temperature as well as pressure, as they are much sensitive to these physical conditions. By utilizing inorganic nanoparticles, a robust antibacterial activity can be attained in low concentrations and high stability in extreme conditions.

Hence, in recent few years, it has been a major interest of utilizing nanoparticles in the packaging of antimicrobial food. An antimicrobial kind of packaging is actually a kind of active packaging that contacts with the food or headspace inside so as to inhibit or retard the microbial growth that might be available on the surface of food (Soares et al., 2009).

Various nanoparticles for example copper, silver, chitosan, and metal oxide nanoparticles such as zinc oxide or titanium oxide have been conveyed to have antibacterial characteristic (Bradley et al., 2011; Tan et al., 2013)

The use of nanoparticles is not only restricted to the antimicrobial food packaging, but nanolaminates and nanocomposite have actively been used in food packaging to provide a barrier from extreme mechanical and thermal shock increasing the shelf life of food. In such a way, the integration of nanoparticles into materials of packaging provides food of good quality along with the long shelf-life.

The main aim of making polymer composites is to have more thermostable as well as mechanical packing materials. Various organic and inorganic fillers are being utilized so as to attain improved quality of polymer composites. The integration of nanoparticles in different polymers has permitted developing much more hard packaging material with the effectiveness of cost (Sorrentino et al., 2007). Application of inert nanoscale filters like silicate and clay nanoplatelets, chitin or chitosan, silica nanoparticles into the polymer matrix renders it stronger, lighter, fire resistant and better thermal characteristics (Duncan, 2011; Othman, 2014).

The antimicrobial nanocomposite films that are prepared by infusing the fillers into the polymers provide two-way advantages due to their barrier properties as well as structural integrity. These films have one dimension in the nanometric range.

3.4.2. Nanosensors for Pathogen Detection

Nanomaterials for application in the building of biosensors provide the high level of sensitivity and various other ideal attributes. In food microbiology, Nano-biosensors or Nanosensors are utilized for the pathogen detection in processing plants or also in food material, changing consumers and distributors on the safety issue of food and quantification of existing food constituents (Cheng et al., 2006; Helmke and Minerick, 2006).

The Nanosensor functions as an indicator that reverts to alterations in the conditions of environment like temperature or humidity in storage rooms, product degradation or microbial contamination (Bouwmeester et al., 2009).

Different nanostructures such as nanofibers, nanoparticles, nanorods, thin films have been analyzed to their possible use in biosensors (Jianrong et al., 2004). Optical immunosensors based on thin film, for the detection of microbial cells or substances, have led to the fast as well as much sensitive detection systems.

In these kinds of immunosensors, particular antigens, antibodies or various protein molecules are immobilized on thin Nanofilms or sensor chips that produce signals on detection of target molecules (Subramanian, 2006).

A dimethyl siloxane microfluidic immunosensor incorporated with particular antibody that is immobilized on an alumina nanoporous membrane was developed for the fast detection of foodborne pathogens such as *Staphylococcus aureus* and *Escherichia coli* O157:H7 with electrochemical impedance spectrum (Tan et al., 2011).

Detection of pesticides, toxins, and pathogens can also be done by using techniques of nanotechnology (Liu et al., 2008; Inbaraj and Chen, 2015; Palchetti and Mascini, 2008). Different biosensor that are based on carbon nanotubes have achieved so much attention because of their fast detection of toxins, microorganisms, and various other degraded products in food and beverages (Nachay, 2007).

The attributes of toxins that are attached to these nanotubes cause a change in conductivity that is detectable when destined with numerous waterborne toxins and hence, are utilized to detect waterborne toxins (Wang et al., 2009).

Additionally, the application of electronic nose or tongue that are consists of the series of Nanosensors monitors the conditions of food by giving distinct signals on gases or aroma released by food items (Garcia et al., 2006).

The quartz crystal microbalance (QCM) is based electronic nose can easily detect the interaction occurring between different chemicals and odorants that have been layered on the crystal surface of the QCM.

Various studies on the detection of small molecules have utilized quartz crystal surfaces that have been earlier improved with distinct functional groups or biological molecules like enzymes, amines, lipids, and numerous other polymers (Kanazawa and Cho, 2009) (Figure 3.3).

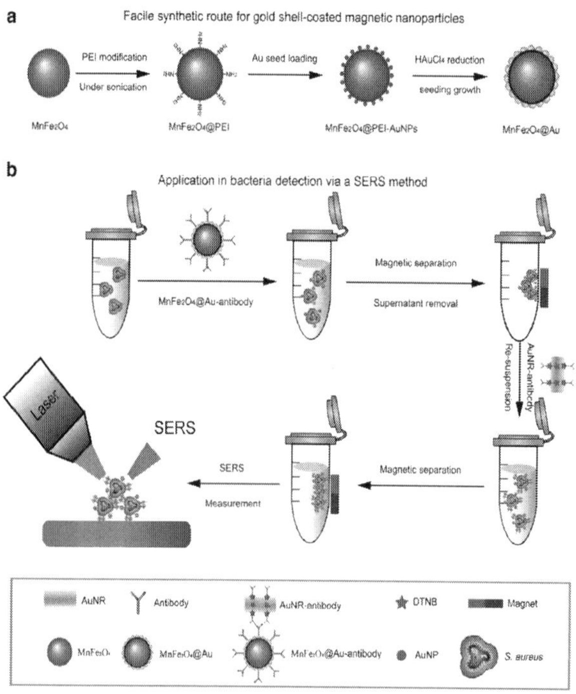

Figure 3.3: Development of nanoparticle-based optical sensors for pathogenic bacterial detection.

Source: https://jnanobiotechnology.biomedcentral.com/articles/10.1186/s12951-017-0260-y

3.5. NANOTECHNOLOGY IN AEROSPACE MATERIALS

There are very few numbers of industries where the use of nanotechnology is clearly advantageous as in the industry of aerospace. The key development goals match exactly with the benefits provided by applying different nanomaterials in the place of conventional bulk metals such as steel (Figure 3.4).

Figure 3.4: Nanotechnology in aerospace.

Source: https://www.azonano.com/article.aspx?ArticleID=3103

The aerospace industry is one of the most significant heavy industries all around the world. Innumerable companies depend on the capability of ship products and people around the globe with the speed that can be attained by air only.

The market of aircraft manufacturing is very much worth and the bulk of this is accounted by the military spending. Together with this big economic value comes massive consumption and one of the huge carbon footprints on the planet comparative to the market size.

Because of this reason, the key drivers in existing aerospace research and development are in the direction of light construction materials as well as much efficient engines, the major aim being to decrease the consumption of fuel and emission of carbon linked with air freight and air travel. For the industry of aerospace, the important interest in nanotechnology is acceptable by the strength of nanomaterials as well as nanoengineering to aid the industry attain this aim.

3.5.1. Nanostructures Metals

A lot of materials with some nanoscale structures are broadly utilized in the manufacturing of aircraft. Now, it is very well known that nanostructures metals display enhanced characteristics as compared to their counterparts with larger or microscale structure.

Mainly, this is noticeable for the features that are critical for those materials which are generally used in aircraft manufacturing; chiefly produce corrosion resistance, strength, and tensile strength, linked with low density that helps in keeping the total weight of the aircraft down.

3.5.2. Polymer Nanocomposites

Numerous nanomaterials have been utilized as filler materials so as to improve the characteristics of structural as well as non-structural polymers used in the construction of aircraft. The commonly used nanomaterials comprise carbon nanotubes, Nano clays, graphene, and Nanofibers.

Particularly, carbon nanotubes have been depicted to give exceptional benefits when utilized as fillers in different polymers because of their excellent toughness, stiffness, and exclusive electrical properties.

Typically, the nanocomposites have outstanding weight to strength ratios, and increased resilience to fire and vibration, making them supreme for application in the aviation industry. The characteristics of the nanofillers, for example, the nanotube conductivity can develop exciting opportunities for various multifunctional materials.

The characteristics of polymers increased by nanomaterial fillers are very well-tuned to the needs of aircraft manufacturers, that they are really being utilized to substitute few of the metals used in the airplanes. Obviously, this brings along big weight savings as well as cost savings.

3.5.3. Tribological and Anti-Corrosion Coatings

Another key trend in the materials that are used in the aircraft is towards all Nanocoatings to increase the strength as well as life of the metals. Particularly, magnesium alloys that are much lighter than aluminum or steel are susceptible to corrosion because of the high chemical reactivity of magnesium. Coatings can be used to avoid corrosion.

The type of coating that is used typically contains complexes of chromium that are a highly toxic pollutant. Different materials that are used for these kinds of novel anti-corrosion Nanocoatings consist of boron and silicon oxides and cobalt-phosphorus nanocrystals.

Currently, Nanocoatings are being majorly used on the blades of turbine and various other mechanical components that have to survive friction wear as well as high temperatures. The tribological coatings can extremely lower the friction coefficient and enhances the resistance to wear; eventually, this increases the efficiency of the engines.

Various nanoscales and nanostructures coating materials have been recommended as possible friction modifying agents like nitrides, carbides, metals, and various other ceramics.

3.6. NANOTECHNOLOGY IN CONSTRUCTION

The industries related to construction were the only one to recognize technology as a capable evolving technology in the UK Delphi survey in the early 1990s. The status of nanotechnology was emphasized in the foresight reports of the UK and Swedish construction.

Additionally, ready-mix concrete, as well as concrete products, were recognized as amongst the top forty industrial sectors probably to be affected by nanotechnology in coming ten or fifteen years. Although construction has lagged behind various other industrial sectors like chemicals, automobile, biotech, and electronics sectors, the nanotechnology research and development has fascinated substantial interest as well as investment from huge venture capitalists and industrial corporations. Identifying the major importance and potential of nanotechnology to the construction industry, the European Commission at the end of 2002 permitted funding for the Growth Project GMA1-2002-72160 "NANOCONEX" - In the direction of the setting up of a Network of Excellence in Nanotechnology in Construction. One of the key tasks of this twelve-month project was to make a report on the state of the art of nanotechnology in construction (Figure 3.5).

Figure 3.5: Application of nanotechnology in construction.

Source: https://www.slideshare.net/GrimGod/application-of-nanotechnology-in-civil-engineering-46671949

The major goal of the state-of-the-art report was to assess existing awareness, development along with the future potential of nanotechnology in the wide field of construction as well as the built environment. Also, it is aimed at denoting to a long-term map of the path of nanotechnology in construction and providing support to the setting up of various scientific programs of the future proposals.

This state-of-the-art report was made basically in two parts. The first part was completely based on the results of an email survey of the leading researchers as well as professionals who are involved in the construction industry. The second part was a kind of desk study of the global activities that are related to nanotechnology and development related to construction and the built environment by evaluating the current data, reports, publications, and information present on the Internet.

3.6.1. Awareness and Existing Nano-Related Activities in Construction

An e-mail survey in the form of e-mail and questionnaire was performed in two major stages. In the first stage of the survey, a very simple questionnaire was sent to the population of approximately four hundred leading researchers and construction professionals who are in the fields of construction, big companies that involved in construction as well as the manufacturers of the Nanotools or Nanomaterials.

In the second stage of the survey, two additional questionnaires were sent to respondents selected from the first stage of the survey. These respondents were then asked to give detailed information on the current resources and appropriate current research projects.

With the European project in mind, almost 90 percent of the targeted population was from the European countries and the rest 10 percent were from various other countries such as Canada, US, Japan, Australia, and China.

Amongst the total targeted population who received the questionnaire, eighty-three responded back with a return rate of 21 percent. Because of just a few responses from outer areas of Europe, the result of survey is likely to reflect the current scenario of Europe.

The majority of the respondents were from the research institutes or universities, and just around 11 percent were from a huge number of construction sectors different from equipment and material supplier including design, service, survey, contracting, project management, etc.

Few of the response appeared to recommend that the low rate of the industrial response was majorly because of the very limited awareness and knowledge regarding the development of nanotechnology and a lack of interest in being involved amongst the construction firms.

Then it was found that the awareness, as well as knowledge related to the nanotechnology use in construction, was usually very higher amongst the academics or researchers as compared to the industrial personnel.

Understanding the process at nanoscale was the only application or activity in the field of construction known to the majority of the industrial personnel who responded in the survey. Additionally, the fields of activities that the respondents were much more aware of are those of the fundamental research and development materials. The results that were obtained from the survey of professionals, as well as leading researchers those were involved in the field of construction on their contribution in Nano related research and development, was so much advantageous that further development in nanotechnology can be done on the basis of this result.

The results obtained were observed to show that understanding process at nanoscale and Nanoparticles, filler as well as admixtures were the two major applications or activities in construction known to the respondents.

The key activities of the respondents are in the fields of high-performance structural materials and understanding process at nanoscale. It is also observed that except for the restricted contribution in the field of high-performance structural materials, the industrial personnel have very less involvement in the activities that are related to nanotechnology.

Another significant finding coming out of the survey was the attitude and the perception of the construction industry on the developments of nanotechnology along with the potential use in construction. Among all of the respondents, only two agreed that the common people were not much informed or was quite confused regarding the notion of nanotechnology and the condition was worst in the industry of construction.

Mainly, as indicated by few respondents, the information related to the applications of nanotechnology is not at all valued in the construction industry. Nanotechnology was earlier perceived as quite expensive as well as very complex to explain to clients who desire a structure built as cheap as possible.

This distinct with the quick increase in the knowledge and awareness of as well as interest in the development of nanotechnology in various other industrial sectors like energy, automobiles, and chemicals. In replying to the second stage of the survey, all of the 54 selected respondents provided data on their current and existing research projects that are relevant to nanotechnology in construction.

The outcomes of the survey were unexpectedly positive, as around hundred research projects were recognized as relevant to nanotechnology in construction amongst the respondents. Various kinds of research projects are being carried out by the researchers and the leading professional in the field of nanotechnology.

Few examples are high-performance structural materials- nanostructure alteration of ceramics or glass, steel or metals, composites by process control or using nanoparticles, cement or concrete, polymers, nanotubes, and Nano-admixtures; understanding phenomena at nanoscale- macro-

and nanostructure characteristic relationship like shrinkage, hydration, deterioration, etc., by using advanced techniques; functional thin films or coatings- much improved performance in thermal, optical, abrasion resistant, durable, anti-graffiti, self-cleaning, etc.; new sensors, devices, and smart structures- self-actuating capability and improved monitoring of structural as well as environmental conditions, but this inclined to be at micro level instead of macro level; Sustainable energy and environment applications – new fuel cells, special insulating and glazing, energy-efficient lightings, cleaner, pollutants, etc.

3.7. NANOTECHNOLOGY IN MEDICINE

The application of nanotechnology in medicine provides some exciting opportunities. Few of the techniques are just imagined whereas others are at different levels of testing or in fact are being used today. Nanotechnology in medicine comprise use of nanoparticles that are currently under the development phase as well as extensive range research that comprises the application of manufactured Nanorobots to make repairs at the cellular level, also referred to as nanomedicine.

The application of nanotechnology in the field of medicine can revolutionize the method of damage detection and treatment of the human body as well as disease in the coming future, and various techniques just imagined few years back are making amazing progress on the way to become realities (Figure 3.6).

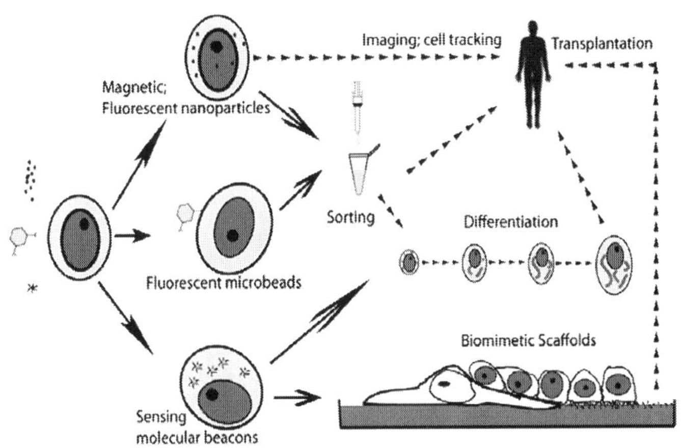

Figure 3.6: Nanotechnology applications in stem cell biology and medicine.

Source: https://www.omicsonline.org/open-access/nanotechnology-and-its-applications-in-medicine-2161-0444-1000247.php?aid=41535

3.7.1. Nanotechnology in Medicine Application: Drug Delivery

One of the major uses of nanotechnology in medicine that is currently being developed includes the application of nanoparticles to deliver drugs, light, heat, and various other substances to the specific kind of cells for example cancer cells.

Particles are also engineered so that they are attracted towards the cells that are diseased, which permits direct treatment of those diseased cells. This technique decreases the damage to healthy cells present in the body and permits for earlier detection of different diseases.

For instance, nanoparticles that transport drugs of chemotherapy specifically to the cancer cells are still under development phase. Tests are in development stage for targeted transport of the chemotherapy drugs and their application with cancer patient is still pending.

CytImmune is a company, which has published the consequences of the Phase 1 Clinical Trial of their very first targeted chemotherapy drug. Another company named as BIND Biosciences has published initial consequences of a Phase 1 Clinical Trial for their very first targeted chemotherapy drug and now the company is proceeding with a Phase 2 Clinical Trial.

At Georgia State University, researchers are using nanoparticles in making an influenza vaccine that targets a part of the virus which is present in every influenza virus. Their aim is to make a vaccine that will effectively work on all of the influenza viruses.

At the Wyss Institute, researchers are testing those nanoparticles that release some kind of drug when forced to sheer force, like that occur when transporting by a part of artery which is majorly blocked by a clot. Various lab tests done on animals demonstrated that this approach is very much effective in delivering drugs that can be further used to dissolve clots.

At the Houston Methodist Research Institute, researchers have demonstrated a method of targeted drug delivery in mice by using silicon nanoparticles that degrade in a tumor and thus release numerous polymer strands that form a nanoparticle having the drug to be transported. This polymer nanoparticle gets dissolved in the cancer cell, transporting the drug to the cancer cell. At the University of Illinois, researchers have demonstrated that the nanoparticles of gelatin can be utilized to transfer drugs to the tissue of brain that is damaged, more effectively as compared to other standard methods. This approach is demonstrated in the lab only, researcher's hope that this approach will become as the most efficient method for drug deliveries for brain injuries.

At MIT, researchers are examining the application of nanoparticles to transfer or deliver vaccine. The nanoparticles basically protect the vaccine, permitting the vaccine time to stimulate immune response as depicted in the lab test with mice. To adapt these kinds of techniques, further research and development needs to be done.

Currently, researchers are developing an approach to release insulin that utilizes a sponge-like matrix that has insulin along with the Nanocapsules consisting of an enzyme. When the level of glucose increases, hydrogen ions are released from the Nanocapsules that bind to the fibers and make up the matrix.

Due to release of hydrogen ions, the fibers become positively charged. The hydrogen ions repel each-other and create openings in the matrix by which insulin is also released. In tests with the lab mice, this approach has been proved to be very much effective.

Researchers are developing that kind of nanoparticle which can be consumed orally and pass by the lining of the intestines to the bloodstream. This must allow drugs that now be delivered with just a shot consumed in a form of pill.

To defeat viruses, researchers are developing a specific nanoparticle also. The nanoparticle does not necessarily damage the molecules of virus but delivers an enzyme that stop the reproduction of virus molecules in the bloodstream of patients. The efficiency of this method has been verified in lab tests.

3.7.2. Nanotechnology in Medicine Application: Therapy Techniques

Researchers have recently developed Nanosponges that are used to absorb toxins and then remove them from the bloodstream of the patient. The Nanosponges are the kind of polymer nanoparticles that are coated with a red blood cell membrane.

This red blood cell membrane permits these Nanosponges to travel easily in the bloodstream of the patient and hence attracts the toxins.

Researchers have established a method to produce sound waves that are much powerful but tightly focused, which may be utilized for noninvasive surgery eventually. They use a lens that is coated with the nanotube made up of carbon so as to convert light emitting from a laser to focused sound waves. The aim is to establish a method that can blast the tumor or any other diseased areas without destructing healthy tissues.

Researchers are examining the application of bismuth nanoparticles to concentrate numerous radiations that are used in radiation therapy in order to treat cancer tumors. Primary results show that the bismuth nanoparticles will enhance the dose of radiations to the tumor by almost 90 percent.

Nanoparticles that are made up of polyethylene glycol-hydrophilic carbon clusters (PEG-HCC) have been demonstrated to absorb free radicals at a very high rate as compared to the proteins out body utilizes for this function. This capability to absorb free radicals may decrease the damage or harm which is caused by the release of free radicals just after a brain injury.

To destruct the tumors formed in breast cancers, targeted heat therapy is now being developed. In this method of targeted heat therapy, antibodies that are potentially attracted to proteins formed in only one kind of breast cancer cell are stacked to nanotubes, causing the nanotubes to gather at the tumor. Infrared light from the laser is basically absorbed by the nanotubes and generates heat which ultimately burn up the tumor.

3.7.3. Nanotechnology in Medicine Application: Diagnostic Techniques

At Worcester Polytechnic Institute are utilizing different antibodies that are attached to carbon nanotubes in chips so as to recognize the cancer cells present in the bloodstream. The researcher believes that this approach can be simply used in labs which eventually will provide early detection of cancer cells if present in the bloodstream.

At MIT, researchers have developed a kind of sensor by using carbon nanotubes that are embedded in a gel, which can then be injected under the skin so as to examine the amount of nitric oxide present in the bloodstream.

The amount of nitric oxide is very much important as it specifies inflammation, permitting informal monitoring of all of the inflammatory diseases. While the testing with mice in the lab, the sensor remained active for almost a year.

At the University of Michigan, researchers are developing a kind of sensor that can easily detect a low level of cancer cells, as low as three to five cancer cells in just a one milliliter of the blood sample. They grow numerous sheets of graphene oxide onto which they assign molecules having an antibody that attaches to the cells of cancer.

Then they tag these cancer cells with the fluorescent molecules so as to make these cancer cells stand out in a microscope. Various researchers have verified an approach to use nanoparticles for primary diagnosis of the infectious disease.

The nanoparticles get attached to the molecules present in the bloodstream indicating the beginning of an infection. When the sample is allowed for scanning, for Raman scattering, the nanoparticles improves the Raman signal, permitting detection of the molecules depicting an infectious disease at an early stage.

For early detection of the damage of kidney, a test is being developed. The method utilizes gold nanorods which are functionalized to get attached to the kind of protein produced by the damaged kidneys. When the proteins gather on the nanorod, the color of the nanorod changes. The test is designed so as to be done rapidly as well as inexpensively for initial detection of the tumor.

3.7.4. Nanotechnology in Medicine Application: Anti-Microbial Techniques

At the University of Houston, researchers are developing a method to kill bacteria by applying infrared light along with the gold nanoparticles. By this method, enhanced cleaning of the instruments present in hospital settings can be observed.

Researchers at the University of Colorado Boulder are examining the application of quantum dots to treat those infections that are antibiotic resistant. At the University of New South Wales, researchers are analyzing the application of iron oxide nanoparticles that are coated with polymer to treat those infections which are caused by chronic bacteria.

One of the initial applications of the nanomedicine was the utilization of nanocrystalline silver as a kind of antimicrobial agent for the treatment of wounds. A cream made up of nanoparticle has been shown to battle with staph infections.

Nitric oxide gas is present in the nanoparticles, which is also known for killing bacteria. Studies done on mice have demonstrated that applying this nanoparticle cream to release gas of nitric oxide at the location of staph abscesses substantially decreased the infection.

Burn dressing is coated with the Nanocapsules having antibiotics. If an infection occurs, the harmful bacteria present in the wound causes the Nanocapsules to release the antibiotics. This permits rapid treatment of an infection and hence decreases the number of times a dressing has to be changed.

3.8. OPPORTUNITIES FOR NANOTECHNOLOGY

Developments in nanotechnology are basically leading to the advancement of cleaner technologies as well as products with better engineering performance. This means that nanotechnology can offer various solutions to numerous issues and challenges related to daily lives of human beings.

A question arises here that how could nanotechnology help the manufacturing industry to become much more sustainable? Basically, there are two methods to look at sustainability and nanotechnology. First is that manufacturers could utilize nanotechnology so as to develop those kinds of products that optimize the usage of resources.

For example, nanotechnology is leading to the advancement of LEDs (Light Emitting Diodes), which has very high efficiencies in consumption of energies during the operation. Batteries of computers can be developed to the last longer ones and hence using less energy during the improvement of computer performance.

Power transmission systems can be enhanced so as to reduce the loss of energy while transmission. Nanotechnology is also leading to the development of various fuel cells of very high efficiency that may reduce dependency on fossil fuels. Second is that manufacturers can apply nanotechnology so as to green their practices. One method is to install solar panels based on nanotechnology. With this method, the product footprint is decreased since the input energy has very few environmental impacts as compared to the fossil fuel. For instance, since 81% of the energy utilized in the life cycle of a desktop is attributable to the processing of materials. Other methods comprise using light emitting diodes in spite of glowing bulbs and utilizing bottom-up manufacturing to diminish material waste.

3.9. SUSTAINABLE MANUFACTURING CHALLENGES

In the 21st century, the industries involved in manufacturing should struggle for transparency as well as economic growth. A method to achieve this is by offering customers with products that meet their requirements while existing in the harmony with the environment. In this method, various issues and challenges will rise that hamper or add trouble in meeting with these needs; these are known as challenges of the SM.

Driving factors for these kinds of challenges comprise growth of population that applies a major burden on the environment by increasing the demand of energy, products, globalization, energy cost and climate change. A strategy to get rid of these driving factors, as well as challenges, is to follow cleaner production.

Cleaner production intends to optimize the use of resources during the stage of manufacturing of the product. A link between clean production and sustainability was proposed by Young et al., (1997). It is as:

Green robust products + Clean production = Sustainable growth

By green products, it must be understood that special care has been taken so as to reduce the impacts during every stage of the life cycle. This consists of handling the societal issues and problems that may arrive during manufacturing of product for example acquaintance of workers to toxic releases.

Robust products will perform effectively as well as efficiently from an engineering point of view. It is still not known that how many products really have any kind of nanotechnology because of the absence of regulation for the manufacturing as well as commercialization of nanoproducts. Hence, analyzing the sustainability of nanotechnologies appears to b a major issue in itself.

The function of nanotechnology in SM is to provide numerous good opportunities so as to deal with the challenges that threaten different industries. Although, there is a requirement to see at the wide picture to analyze all of the factors that require to be considered so as to establish a robust and sustainable universal manufacturing industry.

Factually, the development of nanotechnologies has always profited some of the countries around the globe. Article 15 of the United Nations Declaration on Human Rights states that "benefits from any scientific research and its applications should be shared with society as a whole and within the international community, in particular with developing countries" (UNESCO, 2006).

Considering this statement as a base, it is valuable to review the state of innovation of nanotechnology on the universal scale. A critical factor for the economic growth is intellectual proprietary rights of technology that can be inspected from the perspective of patents.

Patents are considered for promoting as well as establishing the competition which leads to the economic activity in different developed countries and also a consequence in the capacity to disseminate, adapt, and incorporate ideal technologies.

Although, the economic growth in developing countries is never controlled by patents (Bozeman et al., 2003). In the situation that, developing countries can develop effective patent systems, the potential to attain technologies is restricted by a crucial factor that is the cost of patent filing as well as litigation.

Evidently, this condition has an unusual challenge to developing countries and hence, restricting their efforts to attain SD. A probable strategy is for governments to provide numerous incentives to the global manufacturers so as to develop a kind of industrial system that will offer employment along with various other profits to the community.

REFERENCES

1. Bright Hub Engineering, (2018). *Nanotechnology and the Future of Electronics. Part 1-Introduction.* [online] Available at: https://www.brighthubengineering.com/diy-electronics-devices/7761-nanotechnology-applications-for-electronics-part-one-introduction/ (Accessed on 29 March 2019).
2. Diallo, M., Fromer, N., & Jhon, M., (2013). Nanotechnology for sustainable development: Retrospective and outlook. *Journal of Nanoparticle Research,* [online] *15*(11). Available at: https://link.springer.com/article/10.1007/s11051–013–2044–0 (Accessed on 29 March 2019).
3. Rickerby, D., (2014). *Nanotechnology for Sustainable Manufacturing.* [online] CRC Press. Available at: https://www.crcpress.com/Nanotechnology-for-Sustainable-Manufacturing/Rickerby/p/book/9781482214826 (Accessed on 29 March 2019).
4. Rivera, J., Michalek, D., & Sutherland, J., (2014). *The Role of Nanotechnology in Sustainable Manufacturing.* [eBook] p. 6. Available at: https://www.researchgate.net/profile/John_Sutherland5/publication/241026819_THE_ROLE_OF_NANOTECHNOLOGY_IN_SUSTAINABLE_MANUFACTURING/links/00b7d528ecaa11d3ab000000/THE-ROLE-OF-NANOTECHNOLOGY-IN-SUSTAINABLE-MANUFACTURING.pdf (Accessed on 29 March 2019).
5. Sing, T., Shukla, S., Kumar, P., Wahla, V., Bajpai, V., & Rather, I., (2017). Corrigendum: Application of nanotechnology in food science: perception and overview. *Frontiers in Microbiology,* [online] 8. Available at: https://www.ncbi.nlm.nih.gov/pmc/articles/PMC5545585/ (Accessed on 29 March 2019).
6. Soffar, H., (2015). *Benefits and Uses of Nanotechnology in Electronics | Science Online.* [online] Science online. Available at: https://www.online-sciences.com/technology/benefits-and-uses-of-nanotechnology-in-electronics/ (Accessed on 29 March 2019).
7. Soutter, W., (2012). *Nanotechnology in Aerospace Materials.* [online] AZoNano.com. Available at: https://www.azonano.com/article.aspx?ArticleID=3103 (Accessed on 29 March 2019).
8. Understandingnano.com., (n.d.). *Nanotechnology in Medicine.* [online] Available at: http://www.understandingnano.com/medicine.

html (Accessed on 29 March 2019).

9. Zhu, W., Bartos, P., and Porro, A., (2004). *Application of Nanotechnology in Construction.* [eBook] pp. 651–653. Available at: https://www.researchgate.net/profile/Wenzhong_Zhu3/publication/226541329_Application_of_nanotechnology_in_construction/links/542945d90cf238c6ea7d1f73/Application-of-nanotechnology-in-construction.pdf (Accessed on 29 March 2019).

CHAPTER 4

CHALLENGES IN SUSTAINABLE MANUFACTURING

CONTENTS

Every country needs a sustainable environment to work successfully. Manufacturing Companies applies proactive approach for improving the environmental performance. A strong global economy is sustainable only if it integrates social, economic, and environmental well-being. The challenge for the manufacturing companies is to examine the environmental issues and therefore, must develop a vision to control these issues which has the direct and indirect effect on the climate. Concerted and constant efforts are required to meet the challenges resulting from climate change and its effects.

4.1. INTRODUCTION

Firms are growing and developing at a fast rate. There are many trends which have been growing in the market and allowing direction to the firms for implementing sustainability. The firms will be more interested if they analyze the trends which help in making list investments, meaningful contributions and help in creating the value of the company.

Nowadays, environmental, and social externalities have increased which are making an impact on the company's balance sheet. Sustainability is based on the long known factual idea and from manufacturing perspective, sustainability is considered as fundamental ways in the historical trend from skills, physique, flexible, to small lot production. Sustainable Manufacturing (SM) have become more advanced in many fields, and it keeps on changing to find and achieve unexpected cost of production.

There is a paradigm shift required in manufacturing sector from traditional manufacturing model, in which environmental activities in a business were seen only on cost; natural resources are viewed as freely available.

In fact, Climate change and resource reductions are a brilliant example. Climate change has a negative effect on the manufacturing companies, and it also results in an increase in average global temperatures. Before, the companies can produce greenhouse gases and carbon dioxide without any cost. The companies in these days analyze the direct cost of carbon which is revealing itself through tax, cap-and-trade, and other forms of regulation.

Climate Change is likely to result in an increased temperature in the weather, disruption of water ability, rising sea levels, increased frequency, and changing patterns of precipitation and the manufactures may find difficult for developing the greenhouse gases and ultra-efficient resources.

Even the companies face many challenges as they want to earn more profits by providing the proper resources. They also want to think about the environmental benefits as the size of the risk is high. Companies must take a chance in this growing world to increase transparency and improve sustainability.

The financial system has started to look sustainability as a substitution for advanced and sound management. Sustainability Manufacturing makes best use of the advantages offered by properly pricing environmental and social externalities.

Moreover, it is unclear how an optimal future state map would be developed for Micro, Small & Medium Enterprises. However, developing, and maintaining a focus on SM is new and difficult challenge for the countries.

Sustainability manufacturing should be considered as a primary tool for material resources. There are number of organizations which have moved outside legal compliance to environmental management systems and institute management systems. Those companies can achieve and set objectives efficiently associated to water and waste reductions, energy efficiencies, and reuse and recycling programs.

A focus on large amount of data does not reflect the full sustainability of products or services while cost investments continue to be mentioned as factors that drive these improvements. The relationship between sustainability benefit, costs, and return on investment is often weak. In parallel, there was a growth of social accountability, often connected to issues of corporate governance and investor relations. But there are two shifts taking place.

Companies are taking risk for manufacturing the product, and the companies are facing the challenge whether to involve the suppliers or not. The manufacturer may influence the R&D efforts when the companies sustain a proper communication and integration procedure of the supplier which is beneficial for the manufacturer's future.

For Example, in the year 2008, the annual report of Coca Cola Company had resolute that climate change has become a key factor because the company will not be investing and taking the risks in the market.

Competitiveness is a pillar of sustainable production, but many industries follow the market trends so that it will become concerned about the services and performance that meet the needs and costs of supporting those needs.

In Australian report, climate change is the biggest factor in the manufacturing industries as there is high market risk to their business with very less opportunity. The design challenge for manufacturers is to understand better the service performance of customer requirements, environmental impacts, and products.

The challenge for the manufacturing companies is to find solutions for the problems in a socially responsible manner. The industries used to create platform that allow new product service systems to be designed. In addition, to eliminate the wastes, there is need for fundamental shift in business model. The business that first makes the changes will have competitive advantages. Those that don't make efforts won't be a problem because ultimately they won't be around.

The condition of the environment and industry sectors are constantly changing. There is need to discuss the social and environmental consequences of new technologies and govern the arising scientific and social uncertainties and ethical concerns.

Manufacturing enterprises must have accurate information of the market so that they can be more prepared than any other industry sectors. Sustainable development (SD) adopts human activities within the carrying capacity or environmental limits of the planet, now in the future, at all scales from local and global.

The government has alternative options, so the individuals will able to invest in alternative solutions. Corporate leadership may also need to be measured for its ability to involve people at all levels of the organization to address environmental performance. For example, the social acceptability of economic instruments to change behavior – such as water metering and fuel tax – depend on public confidence, which, in turn, depend on the Government's ability to communicate these alternative options. The government policies always come with the solutions and possibilities.

Manufacturing enterprises are concerned about the costs associated with sustainability. The only way to deal with this problem is to address socio-technical systems that span both production and consumption. The role of the industry is to make an impact on the market, and the global manufacturing industry operates at the intersection of energy and materials.

Manufacturing companies should continuously monitor a pipeline of opportunities. Companies that are successful in the transformation are the ones who both react now and capture existing opportunities but also prepare a plan for action to capture the long-term opportunities.

For Example, there is one company who reported that there are more than 50% manufacturers who have already experienced an increase in compliance costs over the past three years. The major challenge for enterprises, government, and industry bodies is to manage the cost consequences of the sustainability agenda.

Ai Group is an organization who is involved in making discussions with the government and gets the help in managing the impacts on the business. The industries which are in various regions of different countries are not able to find the position and struggling to get the position in the new global market. The companies will get tremendous results for the market if they are proactive and innovative to compete or take opportunities to lead.

The new world of business strategies and sustainable technologies is certainly a challenging and exciting emerging reality for the manufacturing industries. Compliance, costs, risks, community expectations, and market competition are the key drivers that will ensure that those who don't adapt will be left behind. The role that manufacturing will play in creating and shaping this world is significant and will require resolute commitment and effective strategies that hold the full amount of sustainable possibilities.

The time has come to think differently with the new ideas and take initiative along with the cooperative efforts, but sometimes, they suffer because of lack of resources and poor information or may be due to financial barriers. Business Planning is important that will satisfy a broad consumer need, and eventuality plan is essential to growth and business success. For example, the Australian Manufacturing Workers' Union (AMWU) is providing a focus and attention on workers and jobs which is most affected in the transition to the new economy.

A strong succession plan creates opportunities for managers, so they must take a proactive approach to reducing their carbon footprint are experiencing surprising gains. There is general indication of strong financial returns, with convincing payback periods, on environmental initiatives and companies are making good use of the supportive skill and funding available.

Manufacturing companies need to expressively gear up to adjust as the challenges that face the world today call for collective action and changes in how businesses and governments function. Non-Government Organizations, individuals, and companies themselves have all taken an interest in the sustainable matters and not only the government who are demanding that manufacturers act in a sustainable manner.

4.2. GLOBAL MANUFACTURING AND THE SUSTAINABILITY CHALLENGE

The industrial revolution who has involved in the new manufacturing processes has brought the instruments and new machines to use the power and resources of earth's natural systems. Manufacturing in the USA is a vital sector which controlled the world in making of finished goods.

Manufacturing firms has achieved a great success, especially in the USA. The biggest challenge for these firms is whether to take the risks or not. They have the responsibility on their shoulders when it comes to addressing the global sustainability challenges. Many companies are facing the biggest challenge due to lack of labor and very fewer natural resources. The concepts and strategies of manufacturing companies are also used by other countries. For Example, in 1980, Japanese manufacturers has challenged and reached U.S. industry in quite a few sectors and involved in quality principles and applied lean manufacturing concepts. Companies that survived the market adjustment caused by the success of Japan established an industrial Darwinism—they adapted to survive in the new global economy.

4.2.1. Survival Skills for a Global Economy

The company which has been successful in learning the new survival skills has followed the same strategy as moving goods and materials by train since the 1800s. Consumers, regulators, and activists demand greater attention to detail regarding social and environmental issues such as climate change and conflict minerals.

Today, the sustainability of the industrialized world has changed, and the world economy is showing concerns which are associated with unnecessary wastes and energy consumption. The manufacturing companies are ready to face the challenges and provide opportunities for the future generations. They are many competitors in the market which has only started to use sustainability principles as a driver for innovation.

For Example, there is one company named Kubota located in Japan and known for its heavy equipment manufacturer who applied more environmentally friendly processes for drying of core coatings and cooling of hot castings.

Many Countries have recognized energy security as a vital concern. The new line resulted in a cycle time of 17.7 hours compared to the previous 132 hours, costs were cut by one third, and carbon dioxide emissions were lowered significantly. Electrolux is a home appliance company located in Sweden which is used to analyze the life-cycle processes as they are ready to produce a more energy-resourceful refrigerator. This refrigerator has beaten the normal regular refrigerator as it also contains more cost than the normal refrigerator. The company is also impressed with this product as they also increase its market share.

Sweden and Japan are the countries which have been growing rapidly in terms of policies and strategies. Global competitors of the company are ready to set the market as the manufacturers are failing to learn new survival skills. They want to achieve closed-loop material flows, promote product stewardship and eliminate hazardous substances. Companies realize that they want new approach to control the issues toward the enterprise, product design, processes, and product end-of-use (EOU) with an eye toward realizing a green and sustainable future.

4.2.2. Sustainability: Corporate Responsibility

Sustainability is a comprehensive approach which is focused on maximizing social and environmental value. The International Organization for Standardization is an association which is responsible for the environmental and the social factors. This organization is focused on the three pillars of sustainability, i.e., economy, society, and the environment. It is a response to the challenges of the modern world facing management from the public and the private sectors.

During the past years, SM has received increased attention by the companies, but there are many challenges which the companies are facing in the economic sector. The companies want more tools and large machines to manufacture the product, so their business will automatically increase. Companies are required more supply chain partners that practice sustainability principles which will help in developing the organization. Green supply chain is related to the environment where the number of suppliers works to improve the social and environmental impacts.

4.2.3. Green is Lean

Lean Production is a cost-effective business processes and requires lot of efforts at addressing manufacturing-related environmental problems. Industry is one of the main sources of reducing or avoids waste totally. For example, less energy-severe processes, alternatives to poisonous materials, and process plans to reduce waste. It is obvious that such methods are wholly constant with the perception of lean manufacturing, and it may be said that green is equal to lean.

Now, certain green manufacturing standards that were once a nice benefit, such as reducing water and energy, minimizing waste and decreasing dangerous emissions, are expected from manufacturers. Manufacturing technologies is not just a social obligation but a growth catalyst but moving towards a green manufacturing has the potential to achieve SD and eradicate poverty on an unprecedented scale with speed and effectiveness.

It has been examined that lean supply chains are not essentially green, and they don't know how to make use of cutting fluids in traditional machining operation such as drilling, boring, and mining. This lack of resilience may increase the risk of supply chain disruption with consequences for manufacturers globally. Industrial hygiene challenges the environmental factors for improving the health, and it has become the important job for the industrial hygienists. They must be able to recognize solutions for the sickness.

Government will also need to consider ways of supporting the greening of manufacturing through institutional support and soft technologies approaches. As investments in energy reduction, safety, pollution, and quality control have show tremendous cost reduction, the investments in green manufacturing will also have its own benefits.

Dry machining and minimum quantity lubrication technologies have become more and more active to avoid potentially hazardous emissions and minimize costs and reduce fluid use. Many companies who are working on the technologies or computer machines have not been able to find the great solution to all manufacturing sustainability challenges. But there are few companies who have completely understand the process during the engineering design process. The US agency has found that lean produces a cultural and effective environment which is highly beneficial to pollution prevention and waste minimization.

4.2.4. Global Trends: Less Waste, Longer Product Life

There is no certain demand for the consumers, but they need the products which are designed according to them. For Example, blacksmiths, tailors, and others in skilled trades are ready to make made-to-order products where they generate their income in the past years.

Manufacturing Companies want the right amount of product in the fastest possible time with a minimum of waste, but they are struggling with the concept of mass customization. They are not able to understand as whether to follow the concept of mass production or the concept of individually customized products or mix the both.

Nowadays, companies are making the strategies for designing the product and recycling waste materials into new materials to generate revenue time to time. The experiment got successful as the product is accepted by the large number of consumers. But in case, if the product will not meet the requirements of the product than it will fall out of use and it should be wasted completely. Different product designs are a new trend which has become popular in the market to match the needs of individuals in the society. As a result, as the trend towards globalization becomes more social, persistent, and cultural differences around the world will drive product designs.

4.2.5. Policies Push Product Recovery

Many manufacturers have followed the concept of robust design, design for assembly and design for manufacturing which has become successful in the past years. Today, the manufacturing companies are following the different concept such as design for disassembly, design for the environment, and design for recycling and more attention is given to these concepts.

These newer concepts are in large quantity being determined by increasing concern for end-of-life (EOL) products as these products does not receive continuing support. Manufacturers want to think again and again related to traditional product life cycles. The new policies which have been adopted by the many companies located in Europe and Japan have the responsibility to handle the EOL products. Many times, the products which is wasted or used are taken by the manufactures to be remanufactured, restored or recycled and avoiding product disposal.

Country like USA is following the same business to recycle the EOL products in order to make profit. Automobile recycling is a good example of the EOL products. The Government adopts a more active approach that the original equipment manufacturers play there in product stewardship. The Government plays a strong role and promised a sector-specific Industrial Strategy which will increase productivity, accelerate the transition to a sustainable economy and drive innovation-led growth.

For Example, In the USA, A robust product recovery sector will require an improved reverse logistics (RLs) infrastructure, enhanced communications among the various value recovery business entities, and technological innovations to support the sector.

4.3. OPPORTUNITIES AND CHALLENGES TO SUSTAINABLE MANUFACTURING (SM) AND CMP

Environmental Sustainability impacts are important to understand, and the society has been facing the environmental challenges over the past years. Companies react differently according to the demand from the customer and market.

Greenhouse gas emissions, environmental impact levels, material availability, and use and energy consumption are all topics of interest. Semiconductor Industry are set for the benefits with the help of newer innovations, but these companies need to regularly innovate so that they get connected with the people.

The biggest challenge for the manufacturing companies is to know how the products should be recycled so that it can make an impact on the environment. The companies are spending more time and energy to find new ways to do things.

Sustainability is transformation and development of the organization whereas sustainable systems and development is acceleration due to the input on the challenges facing the planet from many directions from scientific societies to consumers. It is difficult for the manufacturing companies to deal with the engineers who are completely involved in the production.

It is important to maintain the relationships between environmental, societal, and business aspects of sustainability. The companies are ready for the next generations to design the product with the innovation and advanced technology and efforts are required to maintain these critical relationships (Figure 4.1).

Figure 4.1: Innovation and business models for sustainable manufacturing.

Source: https://commons.wikimedia.org/wiki/File:Business_models_for_Sustainable_Manufacturing.jpg

Sustainability manufacturing is part of a larger concept where the government and the consumers are involved. There always been real risks and some issues to meet the demand of the customers. From the education point of view, there are certain tools and engineering principles which are used by the students to face the challenges in the near future.

Manufacturing Industries account for a significant part of the world's consumption of resources and generation of waste. The biggest challenge for the industries is the energy, and the resources of production are high. Disposal of materials and Costs of control and disposal of materials which are used at the time of production, along with the normal environmental safety and health requirements, are likely to increase as well. Independent of environmental concerns and scarcity of water and other minerals is another risk to industry.

Nowadays, more attention has been paid to innovation and the connection with the earth and economy. The matter is not which holy sphere revolves around the other -- but whether the environment is part of the economy -- or the economy is part of the environment.

Sustainability Manufacturing has no clear definition. Both Government and Industry needs to better understand and determine how to face the sustainable challenges. However, the efforts have been made to improve environmental performance which have move towards thinking in terms of lifecycles and integrated environmental strategies and management systems.

4.4. SUSTAINABLE MANUFACTURING AND GREEN MANUFACTURING

Today, companies have started to accept a larger environmental responsibility that meets the needs of the present without compromising the ability of future generations to meet their own needs. Many manufacturing companies are reusing the product which is safe for the environment, communities as well as customers and it does not harming the environment.

It is not obvious how to apply this directly to manufacturing, but we could consider several categories of "sustainability." These categories depend on whether the resource consumed, or impact, is renewable, or the impact can be accommodated by the environment. There are many environmental problems with the coal; oil and gas as they are removed from the earth that is not considered to be renewable. The greenhouse effect is changing the earth's climate. Every year, large coal-burning plant burns 3 million tons of coal to produce 11 million tons of carbon dioxide. Solar and wind energy would be considered renewable. Impacts can be sustainable if the environment can "absorb" the impact as part of the natural processes.

The carbon dioxide that is released is the cause of greenhouse effect. It has been studied that there is small amount of carbon dioxide present in the air to check the atmosphere ability to accommodate this input. There are fewer resources which are sustainably used but are exceptional.

For example, water is a renewable resource, and it's continuously works its way through the water cycle in a specific area. In manufacturing processes, the usage of water is important in the sustainability, and it has been analyzed that more water is consumed and less is refilled which is taken from the system. In the past years, many liquid product manufacturers has designed and packaged their products in glass, maybe the easiest type of packaging to refill and reuse.

Energy sustainability is defined as the measure of the usefulness of value or quality of a form of energy (thermal or chemical) and measures the "maximum amount of work which can be produced by a system or flow of matter or energy as it comes into stability with a reference environment.

It is helpful in better understanding and recognizes causes of ineffectiveness in the environment. Energy is connected to the environment, and increasing energy efficiency decreases environmental impact by reducing energy losses.

The tools which are continuously affecting the environment are resulted in improving the performance. Many companies are showing their concern about the issue of sustainability. There has been a discussion that how to use the tools that are beneficial to the consumers. The Service companies like consultancy and IT, which offer sustainable solutions to their customers to help the end reduce their environmental footprint.

To make the improvement in many fields has now become the challenging task for the industries. The improvement that reduce resource/material consumption such as water or energy and the impact of the system or process on the environment are greener than other systems and processes and are better.

Green products in manufacturing sectors included those that are low on productions and fuel/energy efficient, deployed renewable energy, used less raw materials, used hybrid materials, are safe (lead-free, eco-friendly dyes, etc.), higher consumer health (fortified packaged food, Sulphur-free sugar), and had SM processes.

4.5. THE CHALLENGES ASSOCIATED WITH BUILDING SUSTAINABLE CITIES

It has been discussed that the main focus should be on the social, environmental, and economic challenges which are related with building sustainable cities. There always been a great challenge for the countries to make things sustainable for the development. Every city government should know that their resources are safe for investment in disaster-proof public infrastructure and renewable resources of energy. According to the survey, there have been fewer opportunities for education in the private sector, and most of them are underemployed. It is has been the responsibility for the government to provide incentives to make their conditions of their house better and provide clean water, sanitation, and drainage to every citizen.

Population is difficult to control, and wealth is increasingly required by people. Population and technology are two different things as urban populations face the challenge of becoming more and more capable in the use of water and energy. The government plays a major part to make cities clean and reducing the generation of waste and improving their recycling systems. The better use of ecological systems and natural resources in the developing cities are said to be high-level cities

For Example, the residents who are living in the various cities like Mexico, New York, Los Angeles, are known as wealthier citizens as they have well- managed their resources and done things according to the system. There has been the great demand of fresh water from these cities as their techniques are different to recycle the water and use generates significant levels of pollution and greenhouse gas emissions at the national and global levels.

As a result, urbanization can be an important supplier to waste generation and high resource use, both with ecological effects at the local, global, and regional levels. The cities are successful in increasing their resources with the help of technology.

The development of smart cities serves two purposes. First, smart infrastructure like smart water meters and electricity grids can reduce usage and costs by raising awareness among individuals about how much they are using, but also by automatically reducing consumption at times of limited demand. In many countries, about one-half of water was wasted until recently due to poor infrastructure; however, after installation of "smart" metering technology the amount of lost water decreased by one-half.

Secondly, smart infrastructure can also improve a city's business climate, environmental sustainability, affordability, and general "livability" as in quality of life.

4.5.1. Convergence, But Greater Vulnerability and Heterogeneity in the Global Economy

Today, globalization offers benefits, challenges, and opportunities to the countries. It can provide tremendous growth for economic growth and country is doing business, so the growth rates have been higher. Quality of life for many people has been improved due to globalization. Many developed countries doing great business to control the environment. Ethical business practices in areas such as product safety, labor, corruption, and environmental stewardship have played a dramatic role in the success and failure of global players.

There are some countries which have developed in terms of welfare of workers, economies and shift their operations to take advantage of lower costs of doing business on other world regions. Industrialized nations are considered as the economic growth of the countries, and the nations like USA, Australia, New Zealand, Japan, and Western European nations have higher per capita Income, advanced technology as well as higher literacy rates.

There are some countries which have not been developed in terms of technology as these nations have low per capita income, low literacy rates and standard of living. Convergence is partly driven by globalization and providing the best possible benefits for the largest number of people.

World economy is becoming more complex which certainly leads to the creation of a world that is more multipolar politically. The sectors which are dealing in manufacturing have noticed an unconditional convergence in labor productivity. Manufacturing sectors have started to make their way to balance the costs and take risks against the potential for gain and Growth. The Manufacturing companies achieve higher levels of revenues for producing the products, or the companies who fail in dealing with the product can achieve a low-cost structure. Structural Transformation describes a reducing support on the primary sector and a shift of resources to higher value adding manufacturing and modern services activities that will remain a necessary condition of sustained economic growth.

These ups and downs will have to be providing somewhere to stay within a global governance rule. Companies are trying to get advantage from globalism by selling products in many countries. The transport sector is an important source for the productions and government has made strong progress to improve the business and economic investment.

Earlier, it has been seen unsustainable consumption patterns in developed countries. The Changing consumption patterns and huge development of global consumption in developing economies will help in providing the strong environmental sustainability challenge. There is a new growth theory on economic development which is continuously making an impact on the global economy. The countries like India and China have shown tremendous improvement as compare to the other countries.

Developing globalization has also increased the recurring interdependence of national economies. Owing to tighter links, they are more helpless to external surprises, and crises like the 2008–2009 global financial crisis spread quickly.

The Economic growth in every country is very different. If the industry is good at making new products, they acquire new productive capabilities. Growth in many countries has slowed down extremely and creating a more dynamic environment going forward looks much more challenging. They lack non-resource dynamic export industries that can generate the energy or decrease in consumer demand with immediate effects can lead to less supply into the market.

Manufacturers can implement globalization for best practices to increase the benefits while qualifying the risks to improve the collective futures. In past Years, Financial globalization or economic globalization have become deeper and very sophisticated when they integrate with world markets as they deal with the country's weakness and the pollution effects arising from such crises.

Country like USA has taken the major steps to give financial support to the workers. As a result, many countries followed the same strategy and learn their lessons, so they make their environment strong. It has been proved that financial globalization has several potential benefits as it also possess new challenges and risks. Global economy indicates that the externalities of national economic policies are increasing and thus require management and better coordination at the global level.

Again, notable first steps in this regard have been taken since the crisis, for example, through the establishment of the Financial Stability Board, and through the issuance of new rules by the Basel Committee on Banking Supervision. There are concerns not only about external shocks but also about heterogeneity in growth experiences, i.e., about the fact that progress is not uniform.

The growth of the countries is not enough as some countries live better on less money, but they remain dependent on other countries to get the resources and technology. Huge cost reductions from low-cost logistics and manufacturing have lowered the cost of living for everyone around the world.

Yet, development strategies based on industrialization and structural transformation following the example of East Asia have become more challenging as rents for simple manufacturing and assembly procedures within global value chains have eroded and as prices for manufacturing goods typically exported by low-income countries have fallen more rapidly with the entry of China into global markets.

It has been predicted that shifting end markets will also lead to shifting patterns of global import demand. The policies are made to solve the coordination failures, and the industries should be competitive enough so that anyone is ready to buy from them. It has observed that too many advance strategies can help in new investment opportunities.

The sustainability indicators are important for the firm's continuous improvement in the environment, and those firms use the crises to end relations with marginal suppliers depend on globally operating suppliers instead.

The government is ready to take the risks and develop infrastructure for the companies. The changes are always for the development of the companies but sometimes these changes will reduce development strategies based on export-led growth, and it has successfully applied in many of the best performing developing countries which will be more difficult in the years.

4.6. SUSTAINABILITY DYNAMICS

SD indicates that future is green energy and it controls the future of global human being. The environmental, social, and economic devices are closely related to each other because it is a masterful balance of meeting the needs of people for future purposes. The Sustainability Dynamics Model consists of three models for growing awareness among the society for sustainable products and services. The contribution that defines the sustainability dynamics model has direct effects on the economy or enterprise as well as on the society and environment (Figure 4.2).

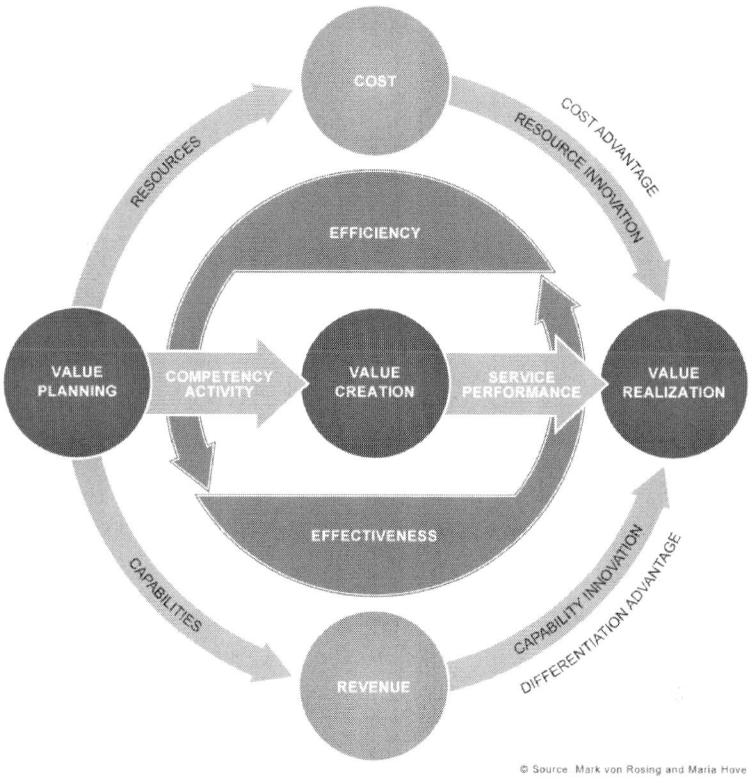

Figure 4.2: Model related to value creation.

Source: https://commons.wikimedia.org/wiki/File:Value_Model.png

Sustainability Dynamics Model has created to examine the consequences of value creation toward environmental, social, and economic sustainability in the future. This model predicts that the social, economic, and environmental goals are different, but they all are related to natural resources.

In the Year 2015, there are approximately 193 countries of the UN General Assembly who adopted the strategy and set sustainable goals to end poverty. The assembly also made changes in the policies for the wealth purposes and for saving the planet.

In the Past, the UN agency has started a movement which is known as project everyone for the purpose of achieving the global goals. It also includes the behavioral changes towards increased sustainability. The movement has started to help and communicate with the people and discuss that how to achieve the goals.

The investment should be made for the production, but it does not harm the planet, and the scope ranges from macro to microeconomic level, from degradation to pollution, from society to individuals along the whole life cycle. It will also result in increasing the quality of life. The opinions of the government and the people such as consumers, policymakers, businessman, and researchers should match to attain the sustainable goals.

Besides, this goal involves a systemic approach and cooperation among stakeholders in the entire supply chain, from producer to final consumer. Every Industry in this century is standing strongly and contributes to SD, but no solution is found easily when seeing the difficulty and complexity of the social systems.

More and more attention required to value creation instruments but many times sustainability often generates a problem between the value of Individuals and value of a whole society.

Sustainable Value Creation is formed by the features of SD and value creation. The mechanisms included in SD and value creation are highly active.

Firstly, Value Creation opportunities for business that embrace sustainability by flows of information, capital, resources, and labor among production systems. Secondly, Value creation improved operational performance and efficiency. Thirdly, value creation runs over two main levels:

1. The micro-economic level manages value creation to improve standard of living and maintain good work-life balance; and
2. The macro-economic level manages value creation of an entire branch and value creation among countries and within regions. It plays a major part in reducing the risks of the business and increase business opportunities. The activities and services involved in the value creation have both direct and indirect impact on the three major dimensions of sustainability such as society, environment, and economy:
    - Causal relations;
    - Magnitude and scale drivers; and

- Latency and timely duration dependencies.

Casual Relations define the resolute effects between a solution, and it has direct and indirect effects on the three dimensions of sustainability. Casual relations in the value creation provide solutions to the society which may be favorable or unfavorable to the people. For Example, a new manufacturing solution has its direct impact on the economy as well as indirect impacts on environment and society.

The direct and indirect impact is determined by the magnitude and scale of a solution's dissemination (e.g., the societal and environmental impact of a solution becomes measurable due to its increasing market share). The effects and impacts have different latencies and time durations (e.g., the societal and environmental impacts of an established solution have a delay and last a certain period of time).

The evaluation and description of these dynamic effects is a scientific task, and its solution has to however be practical at the same time.

4.6.1. Sustainability Dynamics Model

It is a device in which the value creation solutions are the main elements as they are the essential focus of this model. It is established in the middle of the model to investigate the three dimensions of sustainability and provide direct and indirect solutions to the environment, society, and economic sector.

The society, environment, and economy are the dimensions of SD which change around the value creation solution and characterize systems of their own. The starting from the value creation solution has direct effects between the solution, and each sustainability dimension can be pinpointed:

- The Primary effect on the industry is a long-established problem in many countries and concerned with the conversion of energy, greenhouse gases, materials, etc.
- The changes in the lifestyle, products, and wealth have the primary effect on the society. Efficient resources help in building communications.
- The primary effect on the economy is the biggest challenge in manufacturing processes, logistics, factories, etc. The effects on one-dimension system can cause influences on other dimension systems. It's clear that communications in the society will continue to become more strategic to achieve sustainable goals.

In this case, the root causes not only primary impacts on one-dimension system but also secondary impacts on the other two-dimension systems of sustainability. Mutual spiral effects between the sustainability dimensions systems can, furthermore, be caused by the intended primary effects. The effect of a value creation solution on the dimension systems of sustainability can be defined as an inside-out effect. Even so, cause and effect vary with different.

4.7. SUSTAINABLE MANUFACTURING ROUNDTABLE: TURNING CHALLENGES INTO OPPORTUNITIES

Many leading firms are showing a great concern about social and environmental consequences across the world. They are ready for establishing the objectives for improving the economic growth and social equity. The leading firms target is to join sustainability concepts in conventional manufacturing and recognize as well as analyze the drivers in implementing sustainability manufacturing.

The attention is paid by the business managers to understand the concept of SM whereas manufacturing industries produce goods in a sustainable manner. The consequences and challenges which are faced by these industries are societal pressure and public concerns, top-level management involvement and regulations and government policies. These factors are necessary to understand for successful implementation SM.

The global economy is changing and building sustainable culture in organization, and even highly reputable organizations are based on a good core of data and often produce generally different results from one another. For implementing SM, it has been found that societal pressure and public concerns are the most critical driver, but corporate images and benefits are least critical driver for implementing SM.

Culture is incorporeal, but it's hard to measure, address or improve. Even in the organization, sustainability is more of a corporate discipline rather than a corporate function.

The challenges to address sustainability issues are becoming very common in today's world because the current situation is that everybody assumes that natural resources are endless and re-forming capacity of the environment can compensate for all human beings as it is an old way of thinking that can no longer be acceptable.

Countries like India will become the most developing countries in terms of manufacturing the product in all over Asia which will also bring about social issues and considerable environment in the manufacturing as these results in creating lot of opportunities. But companies must stay away from negativity and sustainability will set the tone for a culture of honesty and openness.

SM is a systems approach for the distribution and manufacture of innovative products and services which are used to minimize resources; produces zero waste and removes toxic substances that in effect decreases greenhouse gases across the total life cycle of products and services.

The economic challenges must be faced by the manufacturing companies as they think about the wealth and new services that whether it will give benefit in the future as there so much competition in the market. Environment challenges may come into effect with the use of natural resources. The manufacturing sector is an ever-changing beast and each year the industry faces new challenges. Social challenges related to promoting social development and improved quality of life through new quality of jobs and wealth.

Many countries in the Asia-pacific region are looking into the social, economic, and environmental factors so that the manufacturing companies consider sustainability more seriously. The manufacturing companies is facing increasing regulation and compliance measures as companies need to compete first to market with a new concept.

A clear and fully combined environmental strategy should not only guide skill development; it should also shape the company's relationship to suppliers, customers, other companies, policymakers, and all stakeholders. Companies can and must change the way customers think by creating preferences for products and services consistent with sustainability. The rapid growth in emerging economies cannot sustain in face of mounting environmental deterioration, poverty, and resources depletion.

- **Regulations and Government Policies:** The government often plays an important role to promote a healthy environment for the growth of manufacturing sector, and the government has taken various initiatives by applying policies and procedures for encouraging manufacturing firms to adopt sustainability practices. Every time, the government are coming up with a new industrial policy in all states or regions of the country to stop manufacturing firms from shifting the dirty manufacturing to places with slack environmental legislation. They have also reduced the tax rates for the companies and promote the development of original green technologies in association with financial, technical institutions and help the industry to implement and import established technologies.

- **Top Management Involvement, Commitment, and Support:** Top management are said to be the highest authority in the hierarchy of manufacturing company They are responsible for making the firms sustainable and encourage the firms to evaluate their role in society. They are responsible for deciding the goals and strategies for the manufacturing firms. The top-level management makes sure that the business will move to the right direction along with the clear understanding of this competitive world and develop the policies accordingly.
    - **Holistic View in Manufacturing Systems:** SM needs a full view across in manufacturing processes, product, and supply chain including the manufacturing systems across multiple product life-cycles. There is an interaction between manufacturing and the other functions such as customer involvement in the manufacturing system where the customer is concerned about manufacturing and acquisition costs.
    - **Effective Strategies and Activities towards Socially Responsible Manufacturing:** Policies and Internal strategies are essential to expand internal operations with continuous process improvements and developments associated with sustainability, such as waste reduction, infrastructure development, energy conservation, employee involvement, and emissions control. The firms which have become successful in recent years have successfully adopted the business activities and strategies for increasing the human and natural resources in the future.
    - **Corporate Image and Benefits:** Many of them talk about corporate sustainability which is related to health and safety impacts in a manufacturing firm. It is used for promoting the environmental care so that it can improve the image of the business and increase the manufacturing firm's corporate image. Increasing market competitiveness and creating company's image are measured

as the key driving factors to SM initiatives in a firm. Manufacturing firms that adopt sustainable practices can achieve more profits, increased product quality, and increased market share, etc.

- **Societal Pressure and Public Concerns:** In the past years, there are tough market conditions in many areas such as banks, media, insurance companies, NGOs, and companies don't fully understand drivers and issues. Public concern about societal issues drives the implementation of proactive environmental practices in manufacturing. Natural environment has become a challenging topic for the manufacturing firms that must think about sustainability due to the economic, ecological impacts and increasing awareness of environment safety among society. Companies must transition from meeting regulatory compliance to developing co-advantage.
- **Market Trends:** Manufacturing firms have increasingly started to adjust certain sustainability options within their organizations during these days. Product development in manufacturing industries can become the trend in the market as they are trying to develop and change sustainable products to meet the demands and needs in the market. The business is continuously growing with the market trends.
- **Supplier Participation:** In today's competitive market environment, Companies involve suppliers to play a major role in applying sustainability principles in the supply chain. Sometimes, the selection process is difficult for the manufacturing firms because it effects and extensively participates in an organization's response to concerns about sustainability. The Manufacturing firms should also involve suppliers who represent complexity and high value. A supplier's participation is essential to develop several skills, technologies, and infrastructure, which are mandatory of SM.

4.8. CONCLUSION

The manufacturing sector is a kind of industry that has the capabilities to address the challenges of eco-efficiency, but they need the support from their government to make it sustainable. They fall because they feel that they have lack of structural support it is necessary to meet the sustainability challenge which means addressing greenhouse gas emissions, resource usage, water usage, and the way that these will be influenced by a growing population

Manufacturers are likely to be put in a position because the resources will be scare where they must compete against each other and against the useful resources that could be used for the basic needs of society. The agenda is to capture future value from the more profound sustainability challenges that are projected to arise.

A sustainable organization will take a broad perception of sustainability that includes environmental, social, and economic criteria and involves the entire stakeholder community. Critical elements will be in skills development, new, sustainable technologies and lean and efficient manufacturing practices.

Nowadays, the support is currently lacking, and the sustainability challenges are global, and all countries face similar opportunities and constraints. The challenges which are important in the manufacturing industry are to get clear direction from the highest levels.

The manufacturing industry will be under pressure to become more innovative so that they get to know what they are making and how these products are made, and the modern idea of consumption and economic growth will need to be negotiated.

These challenges will result from the values of the effects of population growth and climate change, with inter-dependent risks from rising global temperatures, regional water scarcity, material resource scarcity, demographic shifts, and mass urbanization.

Manufacturers must consider of new ways to do old things and make old things in entirely new ways to meet these challenges. Companies that fail to take advantage of these opportunities will not be able to give competition in a sustainable global market, and the country will miss out on significant economic growth.

Vehicles and infrastructure are considered as the end products, will have to produce very few pollutants throughout their life cycles and be widely arranged. Some technology to address these challenges is already available, but much has been developed.

The challenge for the manufacturing industries is to capture significant short term and long-term value to hold both radical innovation and efficiency initiatives. Business and Manufacturers may find the ways to distribute the same products with much-reduced impacts and shift to different modes of consumption, including selling performance and service delivery instead of goods.

The new science and technology need to be used to be developed to meet the challenges at different levels and use large amount of energy or generate greenhouse gasses as an essential part of their manufacture. Manufacturing Industry need to develop novel innovations and methods of production in order to meet demanding sustainability targets and create new economic growth.

Manufacturing companies has started their journey as they have been working in improving its practices and continue to give positively change lives so that the business works towards SM.

REFERENCES

1. Nambiar, A., (2010). *Challenges in Sustainable Manufacturing.* [eBook] p. 4. Available at: http://www.iieom.org/paper/Final%20Paper%20for%20PDF/266%20Arun%20Nambiar.pdf (Accessed on 29 March 2019).
2. Stark, R., & Lindow, K., (n.d.). *Sustainability Dynamics.* [eBook] p. 6. Available at: http://file:///C:/Users/lenovo/Downloads/9783319485133-c2%20(1).pdf (Accessed on 29 March 2019).
3. *Sustainable Development Challenges*, (2013). [eBook] p. 4. Available at: https://sustainabledevelopment.un.org/content/documents/2843WESS2013.pdf (Accessed on 29 March 2019).
4. Tennant, M., (2013). *Sustainability and Manufacturing.* [eBook] p. 6. Available at: https://assets.publishing.service.gov.uk/government/uploads/system/uploads/attachment_data/file/283909/ep35-sustainability-and-manufacturing.pdf (Accessed on 29 March 2019).
5. Witlox, G., (n.d.). *Supplier Involvement in New Product Development.* [online] Arno.uvt.nl. Available at: http://arno.uvt.nl/show.cgi?fid=122228 (Accessed on 29 March 2019).

CHAPTER 5

SOLUTION AND IMPLEMENTATION PERSPECTIVE OF SUSTAINABLE MANUFACTURING

CONTENTS

Sustainable Manufacturing (SM) seems to be an easy process for various industries, but it is not exactly the same. There are several challenges attached to SM. But, with every challenge comes a solution. The chapter highlights all those solutions that would help in achieving the right kind of process for SM. It talks about various solutions related to machine tools and for product development. The chapter also outlines the assessment process required for sustainable practices in new product development as this would help in assuring quality and sustainability all together.

5.1. INTRODUCTION

In the event that the ways of life of best in class and furthermore created social orders are molded later on by the current, at present prevailing advancements, at that point the asset utilization at play will surpass each responsible natural, financial, and social limit known to man. The elements of worldwide competition and participation can be used for providing wings to procedures of development and intercession towards a definitive objective and need of maintainability on our globe.

A unique center lies in consolidating building to reasonable assembling, consequently explicitly tending to antique age forming human living. Conceptual, immaterial ideas and objectives overburden individuals, designers, and analysts in various ways. To date, it comprises a mind-boggling assignment to reliably apply a full, adjusted view and basic cross-appraisal of the full scope of important measurements, for example, the earth (incl. atmosphere, assets, and all other normal frameworks), the economy and society—the established three mainstays of the cutting-edge comprehension of maintainability.

Moreover, directing such appraisals on various dimensions of reflections makes an individual sentiment of frailty. Scientists, in any case, go up against the test of researching the two laws of reliance and the hidden center components so as to give new foundational perspectives of the testing ideas and objectives of manageability. Designers, moreover, endeavor to infer strategies, procedures, and innovations to encourage society and organizations in discovering comprehensive, explicit, and proactive answers for manageability.

In that mission, manageability gets separated into controllable components inside a general framework organize: items with their capacities and practices, material determinations, generation frameworks, manufacturing plants, undertakings, strategic components, esteem creation systems, examples of utilization practices, work, and finance frameworks, welfare, wellbeing, etc. The editors of this book and every contributing writer are of the conviction that it is presently high time to give substantial arrangement sets to address different dimensions of "framework driven acknowledgment and conveyance situated" manageability.

However, what does that really involve? In contrast to the fairly broad (however important) discourses encompassing a "total enough set" of manageability improvement objectives, for example, the 17 objectives concurred by the United Nations in September 2015, a basic desperation is joined to chip away at how such objectives should be acknowledged, and which arrangements must be made accessible for them.

Manufacturing and its capability to convey fortune and intelligent solutions for social orders and people has been chosen as a first prime course to examine which changes are essential for achieving a genuine condition of maintainability later on. Feasible assembling in this sense speaks to an assembling designing's way to deal with adapting to these difficulties.

Assembling innovation is produced toward financial aggressiveness, of ecological similarity with characteristic worldwide edge states of asset accessibility and of social welfare with various societal edge conditions to suit the diverse human networks the world over. This multidimensional objective framework can be adjusted by creating sufficient financial, natural, and social criteria, with examination of their interdependencies and use of that investigation for directing mechanical development in particular monetary, ecological, and societal structures.

Before venturing up and taking a stab at a situation from which to set up a "roundabout economy"— an economy where the estimation of items, materials, and assets is kept up in the economy for whatever length of time that conceivable, and the age of waste is limited—it is important to figure out which components should be coordinated into such a "framework circle." Hence, the hidden new mentality of this book expects that the general estimations of manageability and those explicit to reasonable assembling can no longer simply be driven by appraisal elements and objectives of the three measurements condition, economy, and society.

Rather, it has turned out to be important to on a very basic level change center and explicit components and components of the accompanying "interfacing arrangement of frameworks":

- The earth framework with its normal assets and all related environmental, natural, and climatic sub-frameworks.
- The societal system(s) and related personal conduct standards which are exceptionally affected by social, religious, and ethnic qualities.
- The financial system(s) initially determined by a benefit hypothesis dependent on the conventional creation factors, for example, land, work, and capital, reserves which was simply as of late anyway tremendously affected by new plan of action advancements, e.g., caused by the advanced change Research endeavors in this book have explored both explicit specialized methodologies for introducing changes to explicit systems ("the specialized profundity"), and on rather nonexclusive terms, all-encompassing hypotheses and techniques on how esteem creation and its specialized arrangements can be differently affected by explicit earth and financial limit conditions, i.e., the broadness of the general methodology.

The fairly restricted prescient ability of the opportune movement of transformative or progressive changes to the arrangement set of reasonable assembling remains the main nonexclusive test of research in this field. This is the motivation behind why framework dynamic models are esteemed to be suitable possibility for beating such an exploration predicament.

The second nonexclusive test in practical assembling manages the logical inconsistency between:

- The aspiration to examine the diverse parts of the lifecycle conduct of an item as barely as would be prudent; and
- The commitment to give a somewhat lean arrangement of information, data, and advanced models for the structure and assurance of an item arrangement in the "Start of Life" (BOL) period of the lifecycle. The correct blend of both is definitive in the quest for upgrading the likelihood of effect on "shrewd and far-reaching choices" as a component of the designing procedure of adequately supportable items.

5.2. SUSTAINABLE SOLUTIONS FOR MACHINE TOOLS

The assembling business impacts financial, ecological, and social advancement around the world. Industrialized energy intake has been expanding in most established and underdeveloped nations in the course of the most recent years. Almost 33% of overall world's energy utilization and CO2-discharges are identified with the generation business (International Energy Agency, 2007).

Moreover, an undeniably versatile and advanced society is calling for new altered specialized answers for a various cluster of items both whenever and anyplace. In quest for economical change in the assembling business, it is important to create imaginative answers for machine instruments and additionally for generation forms. To meet the requests of the worldwide market, what's more, it's essential to recognize "new courses" for manageable answers for machine apparatus which may serve to create a long haul impact in the production business (Figure 5.1).

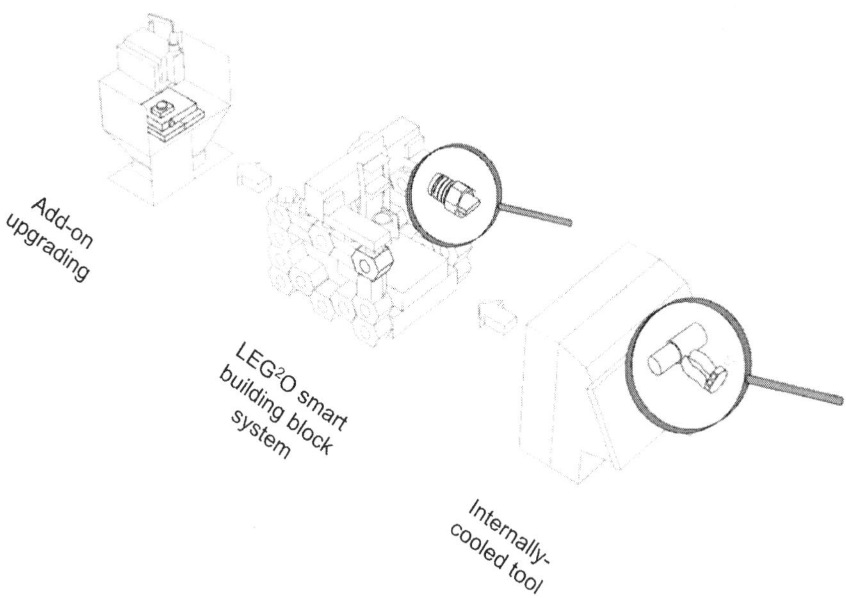

Figure 5.1: Sustainable machine tool concepts- development from the present to the future.

Source: https://link.springer.com/chapter/10.1007/978-3-319-48514-0_4

To affect the maintainability of machine devices, machine instrument outlines, specifically, must establish a focal core interest. Machine apparatuses are "static" as a rule. Current adaptable assembling frameworks can deal with a few generation circumstances. Thus, they are once in a while completely abused and normally "overbuilt" and in that require a ton of designing hours and crude materials. They begin their product life with a negative ecological weight. Even adaptable assembling frameworks are not appropriate for taking care of the cluster measure "one" and in this way, short advancement times are required that can be adjusted to new prerequisites.

These difficulties are the principal drivers for the future advancement of the machine instrument industry, considering progressively rare assets within reach. The expansion in exactness of machine apparatus outlines, the use of mechatronics with respect to the precision of axes, and in addition the "supportable designing" of arrangements and reusing of segments and gear, establish the principal parts of the exploration work displayed in this section.

By applying secluded machine apparatus outlines utilizing microsystem innovation, the exactness of these applications has been expanded fundamentally. Measured machine device edges can be utilized over and over and remain consequently up and coming. Moreover, the acknowledgment of the item clump measure "one" is conceivable whenever required. Using adaptronic segments, the precision of machine instruments is expanded, which means more seasoned machine devices can remain side by side of most recent advancements. These versatile segments can moreover be utilized in particular machine apparatus outlines. By applying inventive apparatuses for machining, feasible answers for machine instruments are in this way being distinguished from a few unique points.

5.2.1. Technological Concepts

1. Microsystems Technology: Microelectronics established the center of the cutting edge worldview "Industry 4.0" and "digital physical frameworks," including supply chains and assembling conditions. Little dispersed frameworks for example wireless sensor nodes (WSN) and -systems (WSS) are for the most part connected inside security frameworks, control frameworks (shut circle administrative frameworks, shut circle supervisory frameworks, open circle control frameworks), observing frameworks (alarming frameworks, data gathering frameworks), e.g., the checking procedure of production gear, yet likewise highlight logistics backing with electronic capacities beyond radio-frequency identification (RFID) (Schischke, 2009).

The ongoing advancement in the exploration of WSN application meets the necessities of assembling situations, for example, utilitarian honesty, vigor, scaling down and low vitality utilization, and additionally the more broad mechanical prerequisites of minimal effort, interoperability, protection from commotion and conjunction, self-arrangement, and association, versatility, information distribution and preparing, asset effective plan, versatile system streamlining, time synchronization, adaptation to internal failure and unwavering quality, application explicit structure, and secure framework structure (Zurawski, 2009).

The utilization of WSN suggests biological conversation starters that should be adjusted effectively with respect to the hard-and programming creators required by methods for tweaked WSN engineering and a WSS design which intently pursues the practical application or utilize case situation while in the meantime keeping up a low ecological impression. WSN for mechanical situations can be sent remotely from the genuine purpose of estimation utilizing complex information procurement systems for measuring, obliged by ecological commotion.

Self-ruling WSNs are connected physically in the fringe condition of the wired framework without direct link get to. Consequently, these frameworks empower detecting errands at every now and again changing or remote areas that can't be gotten to by traditional estimation gear (Ovsthus and Kristensen, 2014).

Fundamental WSN equipment configuration includes a focal preparing unit (microcontroller) that is connected to memory, a correspondence unit (radio frequency (RF) handset including a receiving wire), sensor units, an autonomous vitality source (vitality gatherer or battery) and, alternatively, a human-machine-interface (HMI).

2. **Reconfigurable Machine Tools:** The plan procedure of machine apparatus signifies a noteworthy interest in substantial and elusive assets for machine device maker and buyer. Lead times are particularly high on account of individualized machine devices and mechanical production systems. From a supportability point of view, long lifecycles of machines of over 25 years are hard to oversee, given the unpredictability of item variations and low consignment sizes of the present worldwide market. As the construction of machine apparatuses has high significance on the planet economy, machine device makers along these lines are keen in shortening conveyance times, expanding adaptability and decreasing material utilization in quest for eventually offering better arrangements to clients. Proceeding onward to machine apparatus outlines, these components give principal auxiliary help to each machine instrument.

Their creation requires costly building, testing, and high exactness fabricating. As traditionally threw or welded structures, machine instrument outlines are constrained regarding reconfiguration and can't be changed subsequent to assembling. This limits the reuse to arrangements which were at first fused into the anticipating the piece of the specialists. As the general population is ending up progressively mindful of the issue of manageability, feasible item fabricating has turned into its very own moving purpose right. This is valid for items made with mechanical devices, yet in addition for machine apparatuses as items themselves. Past assembling ideal models went for creating homogenous items at most noteworthy characteristics and least expenses. These days, shoppers are requesting the generation of individualized merchandise. This assembling worldview is called mass customization, which among different improvements, most as of late prompted research on reconfigurable machine apparatuses (RMT) inside reconfigurable assembling frameworks (RMS).

5.2.2. Sustainable Solutions

The ideas and answers for feasible assembling innovation exhibited in Sect. 2 require a well-ordered change inside this control. In this procedure, the initial step to be acknowledged lies in the incorporation of creative machining forms into existing machine instruments. In the second step, the modularization of machine device outlines is to be acknowledged, considering savvy microsystem innovation and inventive machining forms. In the last advance, the overhauling of the machine devices can be attempted as essential. Inside cooled devices turning devices with shut interior cooling frameworks must meet an alternate arrangement of necessities contrasted with traditional devices. Specifically, they should guarantee mechanical security under the temperature scope of the conveyed cooling medium. An inside cooled turning apparatus and cooling outskirts was, besides, created inside the system of the work depicted here.

The extra arrangement grew, besides keeps obsolete machine devices focused in the contemporary creation chain by expanding situating precision as well as by taking into consideration dynamic vibration control to lessen prattle.

- **Smart Building Block Systems:** The innovative premise of the LEG^2O BBS (German abbreviation for lightweight and exactness advanced) comprises of aloof, dynamic, and MST improved shrewd building squares, called modules, with a weight impediment of 30 kg for every unit. In accordance with worker safety regulations for manual lifting, each component of the LEG^2O

- BBS takes into account manual dealing with amid get together, upkeep, and overhauling with low-level framework (Steinberg and Windberg, 2011). Concentrates on utilize instances of these BBS prompted the end that for LEG2 O BBS, most cutting machine instrument situations are sound.
- **Add-Ons for Machine Tool Upgrade:** Obsolete machine devices experience the ill effects of high situating deviations contrasted with current machine instruments and neglect to increase the value of present generation frameworks (Uhlmann and Kianinejad, 2013). Inside the extent of the work depicted here, this specialized out of date quality is tended to by a redesign utilizing an Add-on Accuracy Increasing System (AAIS). The extra frameworks are utilized without trading fundamental framework segments. By overhauling or improving the usefulness of obsolete machine devices instead of supplanting them with new machines, important assets can be spared (Allwood et al., 2011). The extra arrangement created keeps obsolete machine apparatuses focused in the contemporary generation chain by expanding situating precision, as well as by considering dynamic vibration control to decrease prattle the vitality productivity for various machining tasks of an obsolete processing machine and an advanced one was thought about in (Kianinejad et al., 2015). In spite of the fact that the energy utilization of the obsolete machine surpasses that of the more up to date one, the overhaul of an obsolete machine instrument exhibited spares crude materials, vitality for material extraction, alongside assembling vitality by not supplanting the obsolete processing machine with another one.
- **Internally-Cooled Tools:** The instrument plan of turning devices was re-evaluated when taking the prerequisites and states of a coordinated shut circle cooling framework for transforming apparatuses into thought, e.g., cooling medium temperatures from −210 to +40°C. Accordingly, topology streamlined apparatus geometries were reenacted and explored. The subsequent device plan ideas were surveyed by considering the weight, solidness, and the dimension of incorporation (Uhlmann et al., 2014). The situation of the shaper ought to in like manner abstain from moving amid the task of the cooling framework, as this generally results

in stamped varieties in the geometry of the completed parts. To deal with this, the cooling medium channels are coordinated into the turning apparatus holder body and are decoupled from its help structure. Assembling of stream advanced cooling channels is troublesome; consequently, particular laser dissolving is utilized for creation of the apparatus holder.

5.3. SOLUTIONS FOR SUSTAINABLE PRODUCT DEVELOPMENT

Since the mid 1990s academics and practitioners have been placing increasing focus on sustainability mindfulness in the item advancement process, by methods for instruments, strategies, and focused on undertakings towards supportability upgrade. In the early years, the emphasis was on planning better for natural worries, from which period we see the beginnings of what today is a gigantic list of methodologies towards life cycle appraisal and ecodesign, to name only two of the exceptionally prominent ecological enhancement approaches. The critical inquiries are: how can it look today? In which measurements have we built up our insight? How has the world changed since we started to work with ecodesign? What's more, would we say we are viably building up our capabilities, so as to be more powerful in our way to deal with proceeded with manageability improvement? While trying to answer the above inquiries, the creators have done an audit and impression of the past and current decades, previously anticipating our contemplations onto what we see might be the not so distant for maintainability improvement through business-and item advancement (Figure 5.2).

To make this reflection, three eras and nine measurements have been distinguished, in order to portray the general maintainability center, after some time. The eras in center are 1990–2010 (portrayed as the ascent and foundation of ecodesign); 2010–2020 (a frameworks point of view on ecodesign); and 2020–2030 (viewpoints for an economical and Circular Economy). The nine measurements recognized for the audit and reflection practice were the accompanying:

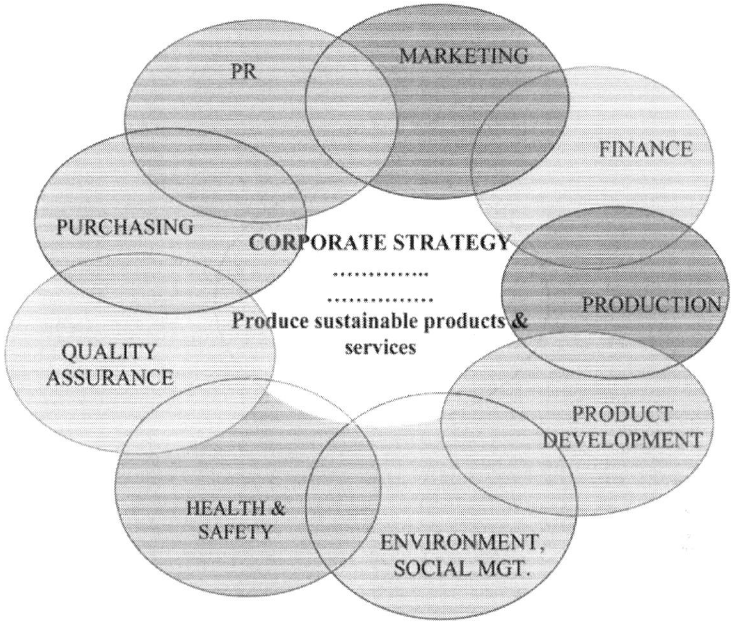

Figure 5.2: Developing sustainable products and services.

Source:https://www.sciencedirect.com/science/article/pii/S0959652602001646

- **Fundamental objectives/goals:** This measurement was incorporated to feature what was the principle supportability configuration protest of the organization, in the given day and age, extending from extremely solid antique centered targets to more subjective destinations found in later occasions
- **Expected outcomes:** This means the fundamental focal point of industry/society in each given day and age, likewise demonstrating the dimension of proactivity towards maintainability inside the given time frame
- **Main point:** This measurement marks whether the fundamental point of the supportability exertion is towards building, actualizing, or completely incorporating instruments into the association
- **Basic methodology:** This measurement helped the creators to separate, regardless of whether the general way to deal with maintainability enhancement could be described as being particular issue centered, framework situated, or all-encompassing

- **Envisaged cost-benefit:** Envisaged money saving advantage: The general mentality of industry, towards maintainability's esteem commitment, was diagrammed in this measurement, to give a genuine picture of the general dimension of desire towards manageability
- **Sustainability ambition:** This measurement signifies which mix of the three supposed mainstays of manageability (ecological social-business) were most in center in the given day and age
- **Business mindset:** This measurement was incorporated into request to separate between officeholder take-make-squander (or 'direct') business outlooks, or whether a more roundabout attitude was obvious in a given day and age.
- **What are we changing:** This measurement was included to put center what the fundamental target of supportability endeavors ordinarily was in a given day and age, regardless of whether it be to make coordinate item enhancements, more foundational process upgrades, or a for the most part all-encompassing spotlight on the abilities of the experts in the item improvement association
- **Decision-production level:** This last measurement was utilized to stamp which overwhelming piece of the association was most instrumentally being locked in, in a given day and age. The accompanying areas survey and think about the exercises, crusades, explore, industry models, and key outcomes picked up from every one of the three separate eras. The above maintainability measurements are utilized as a method for organizing this reflection. A movement and an improvement can be watched, in the three eras considered.

5.3.1. 1990–2010: The Rise and Establishment of Eco-Design

Over the 1990–2010 periods, organizations have fundamentally advanced their methodologies towards the reconciliation of manageability into their business exercises, creating from an uninvolved and responsive position, towards the appropriation of more preventive and proactive methodologies. The business concern identified with manageability issues in this period was straightforwardly identified with the increase of natural mindfulness during the 1980s. The expanded responsiveness was an outcome of the contamination caused by a for the most part uninvolved frame of mind up to that point embraced by industry, where no systems for contamination control were set up. Inside the uninvolved methodology, industrial waste produced in the manufacturing forms by assembling organizations was arranged straightforwardly in the earth with no sort of treatment, prompting an extreme contamination of the natural compartments (soil, air, and water) and making genuine harm both human wellbeing and personal satisfaction. In acknowledgment of the contamination consequences for human wellbeing and nature, governments overall began to escalate their ecological enactment programs during the 1980s, which went for managing organizations' exercises concerning contamination control. From this advancement and reinforcing of natural enactment, organizations began to move from an aloof position towards the appropriation of what we today would call responsive methodologies, which concentrated on the purported 'end-of-pipe' arrangements.

The 'end-of-pipe' arrangements intended at decreasing the contamination capability of mechanical waste, to consent to the authorized enactment, by putting resources into advances, which were mainly expected for the treatment of industrial wastewater, strong waste, and gases created in the generation forms. Because of the generally high ventures for the usage of 'end-of-pipe' arrangements, there was a solid inclination to comprehend natural and manageability issues as an expense to the association, as opposed to as a chance. In the mid-1990s, a preventive methodology rose in a setting in which organizations began to enhance their assembling forms, so as to limit the expanding costs identified with 'end-of-pipe' arrangements, to conform to the consistently obliging enactment and to build asset effectiveness. Ideas, for example, Pollution Prevention and Cleaner Production were important during the time, when the preventive methodology was at its peak. The point was to lessen the waste generation specifically at its source, i.e., in the generation forms, in this manner decreasing treatment and last transfer costs (UNEP, 2004; Ahmed, 2012). Notwithstanding the expenses for the most part ascribed to treatment and transfer, there are different costs that are normally not considered, such as, costs identified with the loss of assets (crude materials, water, vitality, and so forth.), legitimate, and administrative resistance, corporate picture, to give some examples. Ordinarily, for each dollar represented waste treatment or transfer, a further a few dollars are 'covered up' or essentially overlooked, even in very much oversaw and extensive organizations (UNEP, 2004). In spite of the innumerous advantages of responsive and preventive ways to deal with manageability upgrade, only they are insufficient to manage the supportability challenges that our general public was—and still is—

confronting, due to the regularly expanding creation and utilization of items. In the late 1990s, the acknowledgment that items were at the root of the greater part of the contamination and asset exhaustion caused by our general public wound up obvious and a change to a more proactive methodology could be watched. Around then, organizations began to understand that all items caused a type of effect, amid the assembling forms, as well as all through their whole life cycles, from crude material extraction through assembling, utilize, and last transfer (Fava, 1998).

5.3.2. 2010–2020: A Systems Perspective on Eco-Design

In the period 2010–2020 (which embodies the present time of writing), a move can be seen in the public eye, far from the more receptive, device building and particular issue focal point of the primary time. In this period, another influx of globalization is in full stream, empowered by innovation and close moment accessibility of items and administrations, all around the globe. As the world gets less significant, in a manner of speaking, solitary items frequently move toward becoming commoditized, with their apparent esteem lessening to a base. For example, the expanded rate of commoditization can appear to be an immeasurably negative pattern, naturally, due to regularly shortening item lifetimes and substantial episodes of waste, inside a straight economy. Be that as it may, two counter-advancements have risen, in particular, the inserting of high an incentive in great items; and the development of item/benefit frameworks onto the market. High-esteem, amazing items (e.g., premium-evaluated cell phones and top of the line versatile PCs) to be sure give a portion of the response to the past period's concern with commoditization and item squander. Item/benefit frameworks, PSS (which adequately are intentionally co-created item and administration groups) are additionally progressively typical in both B2B and B2C markets. PSS accompany new plans of action, which frequently center around giving more esteem include from one introduced base of an item, by methods for some type of item life augmentation (regularly through sharing), and along these lines dematerialization of the physical antiques, which are segment parts of the PSS under offer (Bey and Mcaloone, 2006). It is in this day and age that numerous organizations are beginning to plan manageability objectives, together with manners by which these will be estimated, be they natural, social or potentially business-situated. The plain escalated time of hardware building has impeded in this decade, with more accentuation being put on the best way to quite choose from the extensive arrangements of devices and techniques and execute the most reasonable instruments inside the organization (Pigosso et al., 2011; Bovea and Perez-Belis, 2012). This is a positive advancement, as we can recognize more than 800 ecodesign best practices as of now (Pigosso et al., 2014) — the spotlight should now be on the most proficient method to guarantee effective usage of these apparatuses and strategies into the business-and item advancement procedures of the endeavor.

Together with the shift from products to PSS as a standard sustainability design object, the basic approach has shifted, so as to incorporate more sustainable decision points at a given time, thereby encompassing a systems approach towards sustainability enhancement. Nevertheless, many companies are not yet realizing the full benefit of their efforts towards sustainability improvement, often rendering sustainability as an activity that may not any longer be seen as a net cost to the company, but is still not a sufficient value-creator in itself. In this decade, social sustainability is a clear focus point for the organization, and a number of projects (often in collaborations between academia and enterprises) have been completed, where social sustainability methods and metrics have been developed, tried, and tested.

5.3.3. 2020–2030: Perspectives for a Sustainable and Circular Economy

A much more critical progress to corporate manageability is normal in the up and coming decade. In spite of the fact that foreseeing what's to come is inconceivable, we have endeavored to build up a situation of how current activities may conceivably send throughout the following decade, in light of an investigation current patterns and past improvements.

Expanding acknowledgement of the need to moderate the impacts of population development, wealth increase and human utilization is presently driving a few worldwide associations to consensually feature the requirement for a noteworthy change in our financial framework, so as to regard planetary limits (Steffen and Stafford Smith, 2013; Häyhä et al., 2016). A few instances of manageability related activities include: the guide for creating vitality productive and low-carbon social orders by 2050, created by the European Union (EU); the 'green development' structure to encourage monetary development while guaranteeing the accessibility of regular assets, by the Organization for Economic Co-task and Development (OECD); and the Sustainable Development Goals (SDGs), propelled by the United Nations in 2016. So as to achieve worldwide and European advancement objectives, the private and administrative segments in Europe need to experience a huge and fundamental change. Because of the acknowledgment of the fundamental supportability challenge looked by our general public, a change towards broadened coordinated effort inside and crosswise over esteem chains is normal. Coordinated effort must be centered on growing new arrangements and monetary frameworks, uniting distinctive partners in the public arena, that assistance tending to the planetary limits (Steffen and Stafford, 2013). An expanding measure of organizations will develop their methodologies towards maintainability and progressively incorporating manageability into not simply the abnormal state vital objectives of the organization, yet in addition the regular business and item advancement forms. This will permit every single choice in the association to be taken dependent on strong and cognizant maintainability contemplations. It will likewise offer ascent to an all-encompassing methodology, in which the associations and interfaces among complex frameworks are considered and their dynamic natures comprehended. Skills will be altogether upgraded to have the capacity to adapt to the comprehension of complex issues and the community improvement of proficient arrangements. Supportability will be characterized and submitted at a vital dimension in associations and the organization into the strategic and operational dimensions will be empowered by the improved development of organizations on manageability upgrade. Now, organizations will have the substance and the setting to have the capacity to comprehend that maintainability squares with business, and that there is no other elective method for being fruitful in a business setting. Truth be told, such signs are now clear in the specific driving edge enterprises, which have put an immediate connection among maintainability and business-improving advancement (Ellen MacArthur Foundation, 2015a). First, on accomplishing a minimum amount of this kind of organization, perceiving the chances of business-driven maintainability activity, will we see that the supportability idea characterized as the harmony between the natural, social, and financial measurements will, at last, be completely met. In the following decade, issues, and dangers identified with asset shortage and item transfer will be limited by an upgraded take-up of the idea of Circular Economy (Ellen Macarthur Foundation et al., 2015), which is as of now being supported in numerous parts of the world. Round Economy is progressively observed as a key way to deal with operationalizing objectives and supporting the change by improving intensity, monetary development and manageability in numerous parts of present-day society. Roundabout Economy is characterized by the Ellen MacArthur Foundation as an economy that gives various esteem creation systems, which are decoupled from the utilization of limited assets.

An effective progress to Circular Economy requires a fundamental change in the manner in which organizations comprehend and work together, with supportability as a solid establishment. Circular Economy will be empowered by the joined use of three segments as below:

- Business Model Innovation;
- Sustainable Design and Ecodesign; and
- Internet of Things coupled with Digital Transformation.

5.4. ASSESSMENT OF SUSTAINABLE PRACTICES IN NEW PRODUCT DEVELOPMENT

Over the most recent few decades, noteworthy research work has been done so as to explore diverse methods for supporting specialists in the advancement of more practical items. In any case, the majority of the endeavors and studies are basically coordinated towards the ecological part of maintainability as the modern world has changed its way to deal with the earth. The significance of the natural supportability of mechanical items and procedures gets not just from the ever stricter-getting to be ecological enactments issued in the vast majority of the created nations, yet in addition from the higher familiarity with clients concerning natural issues. Specifically, the aggressiveness of putting available more maintainable items is turning into a key factor lately. Item advancement is a standout amongst the most basic angles for organizations in achieving their manageability destinations as every one of the items are yields of the item improvement process. Specifically, early plan choices can have an exceptionally noteworthy effect on manageability. These choices concern decisions of material and assembling as well as strongly affect the item's whole lifecycle.In light of the estimation from the EU report, 80–90% of all item related ecological effects are resolved amid the plan period of an item. Thus, eco-plan is an approach to enhance items' life-cycle ecological exhibitions by efficiently coordinating natural perspectives at a beginning period in item structure. In actuality, key difficulties must be defeated to empower ecodesign techniques to be material in early plan stages. With this examination, we mean to comprehend the fundamental inspirations, confinements, and adequacy of incorporating manageability in new item plan and to see how and what organizations are getting along as of now to coordinate maintainability in their items and item advancement forms. In doing this, the emphasis is for the most part on organization attributes, practical systems, and utilization of the organization, ecodesign devices, and significant item advancement forms. In this unique situation, we distinguish the fundamental research goals as underneath:

- Understand how and what organizations are performing to incorporate sustainability into their items and item configurations forms.
- Regulate needs of organizations during item configuration and boosts for supportable item plan.
- Understand the dimension of thought given to reconciliation of maintainability into item plan on the business side.
- Develop understanding into the utilization of eco-structure apparatuses in organizations' item configuration rehearses.
- Identify the hole between the writing and the training.

This chapter is organized as following: State of the workmanship for applicable ideas and the system utilized for planning the study is given to some degree 2. The technique pursued is clarified in detail to some extent 3. Next, the consequences of the examination is appeared and talked about to some extent 4. The fundamental experiences from the paper and ends are introduced to a limited extent 5 lastly section 6 makes reference to the restrictions of the investigation together with a few thoughts for future research.

5.4.1. State of the Art

- **Framework of enablers for integrating sustainability in NPD:** The examination and the study are developed dependent on a structure from a past investigation of the creators. The structure has three principal parts: Drivers/hindrances, the empowering influences which cultivate proficient and viable combination of manageability in NPD (i.e., vital standards, supporting apparatuses and assembling process ideal models) and effect of incorporating supportability in NPD on NPD execution measures (Figure 5.3).

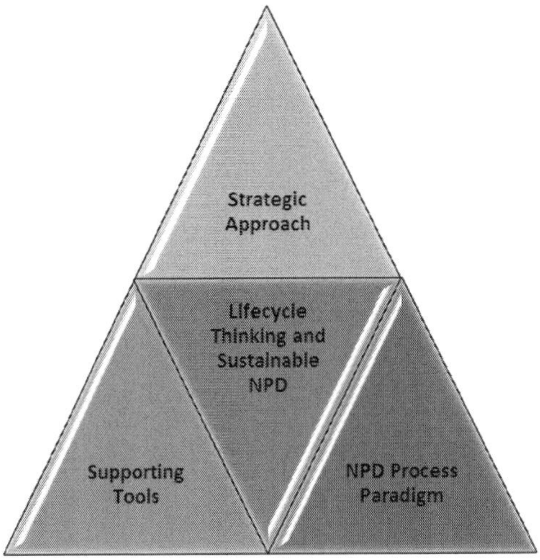

Figure 5.3: Framework of enablers for successful integration of sustainability in NPD.

Source: https://pdfs.semanticscholar.org/7543/e573de4eef2edbd2fe374d04d-55fae7a24dc.pdf

- **NPD process paradigms:** Analyzing NPD process standards concerns surveying how organizations sort out their NPD forms so that supportability is successfully coordinated into item plan. To structure any item, some arrangement of procedures ought to be pursued, and kinds of procedures rely upon the sort of items and additionally organizations' contemplations of various viewpoints and related choices in item improvement.

Thus, the structure of a plan and improvement process influences the supportability execution of an item to be planned. Traditional NPD process is commonly organized in a successive way called sequential designing. Consecutive structure process models are described by solidifying plan particulars from the get-go in a plan procedure with the goal that creators have a solitary plan alternative to recognize till it achieves a control door. At configuration doors, the plan idea will be tried on the off chance that it meets structure necessities, and by then if the structure doesn't meet the prerequisites either structure cycles or problematic item will be the outcome. Subsequently, organizations initially choose the destiny of the item right off the bat all the while and issue will be uncovered later at the structure entryways in such process models.

On the off chance that economical items should be planned after such process worldview, two issues will happen; the first is that, if the originators found that the item does not meet some ecological necessities at last structure doors they have to upgrade once more, which thus influence time to advertise and acquire extra expense for the organization.

Second, since ecological contemplations are taken auxiliary in ventures when natural issues have been found at the entryways, architects neglect the issue and like to dispatch problematic items from supportability points of view.

- **Eco-design concept and tools:** Ecodesign is a training by which ecological contemplations are incorporated into item and process building plan methods. Ecodesign rehearses are intended to grow naturally good items and procedures while looking after item, value, execution, and quality benchmarks. Numerous endeavors have been made to create apparatuses that help incorporation of manageability in NPD.

The most ground-breaking and common devices utilized in maintainable item configuration are those which think about ecological part of supportability, purported "eco-plan apparatuses."

Numerous the alleged eco-plan devices and strategies exist: some are great basic and subjective, (for example, agendas), while some are perplexing and quantitative (Such as LCA), and others dependent on QFD, (for example, Green QFD). The determination of the best apparatus for a given application relies upon the individual circumstance of the setting of the structure and improvement process.

The least complex order of apparatuses is into those which perform information examination, and those, which are gone for development. Investigation devices give an estimation of the potential natural effect of an item. They are for the most part utilized before configuration begins, by dissecting a past item or that of a contender. Then again, they might be utilized toward the finish of a plan undertaking to confirm the outcome.

Enhancement devices, then again, are utilized amid the structure procedure to coordinate movement and give data on the procedure. Slip-ups in choosing the most appropriate device relying upon the explicit circumstance may restrain the viability ease of use and pertinence of the devices. Criteria that ought to be considered amid the determination of the sufficient devices could be: the point of leading the examination, the kind of business or item thought to be, dimension of data accessible, the time accessible, the nature of information input, the nature of the normal outcomes, the proposed client and the plan organize an apparatus is expected to be utilized (i.e., idea stages, framework configuration level, encapsulation structure, and detail plan).

In this examination, we have considered around 30 eco-plan instruments that have been found in the writing. In the review, we researched organizations in the event that they embrace these sorts of apparatuses in their practices and how proficient and compelling they are in applying such devices.

5.4.2. Research Methodology

The examination has been done by methods for a survey created in arrangement with the exploration goals dependent on a system to evaluate manufacturing organizations working in Italy in mechanical, electrical, and automotive segments. The investigation was done on 10 organizations with high development cycles, which are numerous item based EU enactments (i.e., WEEE, ELV, EuP, REACH, RoHS, and others). The survey was upheld by two reciprocal contextual investigations including personal interviews and analyzing the inner reports, to expand the pertinence of the work.

- **Survey design and procedures:** In view of the exploration targets and system depicted in area 2.1 that recognizes basic components to effectively incorporate maintainability and life-cycle thinking in NPD process, a survey of 25 questions has been produced to evaluate the organizations. This study was made out of inquiries concerning organization qualities, maintainable procedures and uses of the organization, ecodesign devices, and applicable item improvement forms.

The survey has been affirmed after some amendment stages in which a portion of the inquiries have been adjusted so as to streamline their comprehension, and different inquiries have been included, so as to increase all the conceivable data lined up with our goals. The normal beneficiaries of the survey were item improvement division dependable.

- **Survey questions:** Survey was depended on the system to actualize manageability in new item improvement (NPD): So, questions have been recognized thinking about the fundamental segments of the structure as recorded underneath:
- Drivers/Barriers:
    a. The upgrades factors considering supportability in NPD.
    b. Barriers of consolidating manageability in NPD.

c. Causes of Eco-structure ventures disappointments and triumphs, etc.

- Empowering influences

  Vital methodology
  - Mandatory and deliberate maintainability arrangements and enactment organization receives.
  - Commitment of best administration.
  - Environmental parameters utilized in item plan.
  - Sustainable item strategy.
  - Expected advantages of applying ecodesign.
  - Trade-off between conventional plan exhibitions (e.g., cost, quality, time to showcase) and environmental and social issues, etc.
- B. Supporting devices
  - Types of supportability configuration instruments utilized.
  - Criteria to pick explicit instruments.
  - The effect of utilizing supportability configuration apparatuses on the NPD exhibitions, and so forth.
- C. NPD Process Paradigms
  - Systematic combination of manageability issues all through a plan procedure.
  - Coordination between item's lifecycle accomplices.
  - Product's lifecycle thought.
  - Exploration and usage of past task's information.
  - Level of advancement received (e.g., item enhancement, item update, new item idea, new item framework, etc.).
- NPD Performances
  - Financial and non-budgetary advantages picked up.
  - Impact of incorporating supportability on the conventional NPD exhibitions, for example, time to market and undertaking costs.
  - Internal changes because of manageability thought in NPD.
- **Sample firms and response collection:** Questionnaire was sent to a mix of expansive organizations and SMEs that may have just been incorporating supportability into their item configuration forms for quite a long while. Sample firms were chosen from assembling organizations working in Italy in mechanical, electric, and car divisions.

Despite the fact that the survey was proposed to more than 200 organizations, we got the reacts from and dissected 10 organizations with high advancement cycles, which are liable to numerous item based EU enactments (i.e., REACH, RoHS, WEEE, EuP, ELV, and others).

    - The respondents were all generally expansive organizations. A portion of the SMEs reacted to sends to educate whether they don't consider this sort of methodologies amid item structure or they are not intrigued by such exercises. Hence, this data may be considered as a sign for a low-level reception of manageable practices and particularly ecodesign apparatuses among SMEs.

- **Complementary case studies:** Two correlative contextual investigations should be conducted to help the outcomes accomplished by the study examination. The point was to gather important data about sustainability practices and maintainable item configuration procedures of the organizations through up close and personal meetings, examination of their inner reports and going further in their reacts to review, as reciprocal to overview investigation.
    - The two organizations chose for contextual analyses were all inclusive known substantial ventures; one car maker and one tire producer. Eye to eye interviews have been made with specialized capable in the two cases. The other component of the investigation was to break down the archives, for example, yearly supportability reports, economical methodology reports, and so on for supporting the outcomes picked up by the earlier overview consider.

5.5. IMPLEMENTATION OF SUSTAINABLE PRODUCTION PRINCIPLES

For the majority of the twentieth century, most makers were centered solely around enhancing their activities to accomplish more effective outcomes. These days, the bar has been set higher, and makers are cognizant about their activities' effects on the triple primary concern or 3BL (individuals, planet, and benefit) and additionally to represent the asset utilization and impression created by their assembling forms (Kleindorfer et al., 2005) (Figure 5.4).

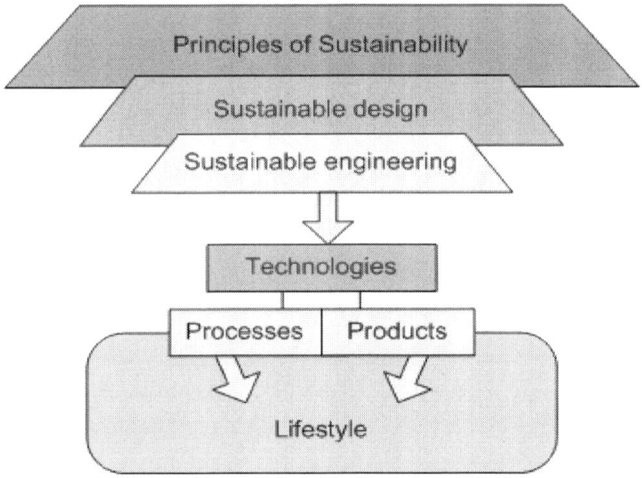

Figure 5.4: Hierarchy of sustainability guidelines and role of technologies emphasized.

Source: https://www.e-education.psu.edu/eme807/node/688

Levels of popularity regarding manageability are thought to be handled by extensive associations, as well as by little and medium measured endeavors (SMEs). Nonetheless, thinks about demonstrate that there is requirement for further examinations identified with manageable practices in SMEs (e.g., Biondi and Iraldo, 2002; Vives, 2005; Lawrence et al., 2006). As indicated by the word reference definition, the expression "standards" suggests settled or foreordained arrangements, method of activity or behaviors. Shrivastava and Berger (2010) characterized standards as "sets of qualities, gauges, rules or decides of conduct that depict the duties or appropriate practices for associations."

For them, supportability standards manage "pushing associations toward manageability by changing their vision/mission, their utilization of regular and HR, their creation and vitality rehearses, and their items and waste administration." This paper is keen on realizing how Swedish makers are as of now executing practical generation standards inside their creation forms. It is relied upon from the associations to draw nearer to an economical creation state when these standards are actualized.

5.5.1. Conceptual Framework

- **Sustainable production:** The LCSP (1998) has characterized feasible generation as "the formation of merchandise and ventures utilizing procedures and frameworks that are non-contaminating; saving of vitality and characteristic assets; financially reasonable; protected and stimulating for representatives, networks, and shoppers; and socially and imaginatively remunerating for all working individuals."

For Manzan and Miyake (2013) the extension considered by the LCSP in characterizing manageable creation is more extensive in correlation with other elective ways to deal with impart natural concerns (i.e., the selection of end-of-pipe advancements for contamination, cleaner generation, and green creation or green assembling). This as it includes the creation of merchandise, the effects on the earth, working condition and social obligation.

As opposed to the previously mentioned models of generation with natural concerns, Manzan, and Miyake (2013) referenced that manageable creation is the special case that advocates for compensating the laborers imagination, having association with the network from a financial, social, and social point of view; and holding agreeable relations with the inventory network.

- **Sustainability standards:** In the wake of completing a short writing audit in Scopus utilizing the watchwords Sustainable creation standards, it was discovered that one of the primary endeavors made to distribute economical standards was made by Daly (1990). He decided the maintainability standards from a high and general point of view.

A portion of the standards proposed by him were: collect rates should rise to recovery rates and waste outflow rates should level with the characteristic assimilative limits of the biological systems; the reasonable utilization of non-renewable necessitates that the interest in the misuse of these assets must be combined with a remunerating interest in an inexhaustible substitute; innovations that expansion asset efficiency must be organized over advancements concentrated on expanding asset throughput.

Moreover, Gladwin et al., (1995) introduced an arrangement of eight operational standards and related strategies of supportable conduct. Veleva and Ellenbecker (2001) presented six fundamental angles to handle when the point is to have supportable activities: Energy and material use, regular habitat, social equity and network advancement, financial execution, laborers, and items. These perspectives are shape dependent on the nine standards of manageable generation proposed by LCSP. For Al-Yousfi's (2004) standards made accentuation in limiting the utilization of nonrenewable assets; decreasing/taking out dangerous/lethal and destructive discharges/squanders, and accomplishing objectives cost-successfully. This as markers can quantify reasonable improvement and maintainable creation at a worldwide, national, sector-wide, or organization levels, or even at numerous dimensions, as the ones given by Veleva and Ellenbecker (2001). Amid our writing survey, we found that an extensive extent of papers drops out of the extent of this investigation. A standout amongst the most well-known purposes behind avoidance was not to manage maintainability standards but rather they utilized the term manageability to make reference to the limit of performing or protecting in the long haul an explicit assembling practice. A few investigations tended to supportability from an open strategy viewpoint, from an explicit industry gathering, others demonstrated administrative activities for creating sets of economical standards, and just a bunch of papers thought about maintainable standards from a generation point of view (shop floor). From this little gathering of papers, just the standards by LCSP were managing the three components of maintainability (condition, individuals, and benefit), while whatever is left of papers were fixated only on the natural element of manageability. At long last, as there is a requirement for inquisitive about manageable generation standards from a triple main concern point of view, the LCSP standards established the best core values for this investigation.

5.6. IMPLEMENTATION PERSPECTIVES

5.6.1. Sustainable Value Creation—From Concept Towards Implementation

We presently reside in a universe of obliged assets, developing populations and surpassing planetary limits. There is a requirement for industry to change the manner in which we make things and move towards a more feasible modern framework. Comprehension of framework change and value change are critical ideas for progressing towards a more sustainable modern framework. Senge (1990) states that the unhealthiness of the present reality is indirectly related to our failure to consider it to be an entirety. Organizations may not be mindful of the full scope of potential esteem results. Most existing plans of action are for the most part dependent on making, conveying, and catching monetary esteem, with constrained or no regard for natural and social esteem. The changing business condition, more extensive scope of partners participating in discussion over industry, asset restrictions and accentuation on social duties of firms has raised the requirement for feasible esteem creation.

5.6.2. Key Concepts for Sustainable Value Creation

1. **Systems thinking:** Systems thinking are a strategy for researching whole frameworks, looking to comprehend the connections, the collaborations, and the limit between parts of framework (Senge et al., 2008; Cabrera and Cabrera, 2015). Systems thinking is especially appropriate to displaying exceedingly complex open-frameworks where a coordinated comprehension is required at both the smaller scale and large scale levels so as to foresee or oversee change. This stands out from the overwhelming expository methodology of the physical sciences, which depends on reductionism, examining shut frameworks at the dimension of their constituent parts and after that streamlining to reach out general inferences. Systems thinking are nonexclusive term that traverses a scope of in excess of 20 apparatuses and techniques (Reynolds and Holwell, 2010). Senge (1990) clarifies that frameworks believing is a control for seeing wholes. It is a system for seeing interrelationships as opposed to things, for seeing examples of progress instead of static depictions. It creates the impression that frameworks believing are a method for moving toward issues: instead of applying a strict straight approach, the strategies are iterative and intended to animate examination, dialog, and discussion by empowering different points of view. Systems thinking do not mean to give quantifiable responses to explicit issues, yet rather gives a scope of choices and better comprehension of the ramifications of those alternatives.
2. **Whole system design:** Whole systems design is one way to deal with feasible plan offering incredible potential; anyway the procedures, standards, and techniques managing the entire framework approach are not obviously characterized or comprehend by rehearsing originators or structure teachers (Charnley et al., 2011). Evans et al., (2009) depicts while it is imperative to address the effect of every part of the modern framework and seek after forceful decrease in the effect of explicit exercises, we should likewise look at the activity of the entire framework. Productively fabricating items that are wasteful being used, for instance, isn't sufficient. This methodology can even

outcome in generously negative results when effectiveness gains or cost decreases result in increments in utilization (the purported Rebound Effect). The best chance to decrease the effect of the modern framework on the planet emerges when we think about the entire framework and the improvement of any individual segment of the mechanical framework. Rough Mountain Institute-RMI (2006) characterize entire framework plan as 'improving parts as well as the whole framework … it takes resourcefulness, instinct, and cooperation. Everything must be considered at the same time and dissected to uncover commonly worthwhile collaborations (cooperative energies) and also bothersome ones.' Entire frameworks masterminds see wholes rather than parts, interrelationships, and examples, instead of individual things and static previews. They look for arrangements that all the while address different issues.

3. **Systems innovation:** It is contended the advancements required for manageable improvement need to move past steady changes. Reasonable advancement requires the change of bigger parts of generation and utilization frameworks (Boons, 2009). Steady (item and process-related) advancements in existing generation and utilization frameworks may prompt further slow upgrades of supportability execution, however, at last, gradual development every now and again does not prompt a universally ideal framework set up in a multi-dimensional creation and utilization framework space. By and large, supportable developments go past normal item and process advancements and are future-arranged. Maintainable advancement goes past eco-advancement since it incorporates social goals and is all the more obviously connected to the all-encompassing and long haul procedure of reasonable improvement for the short and long haul destinations of supportability. Holmes and Smart (2009) depict the requirement for more research in maintainability drove developments and associations.

Adams et al., (2016) presents a model of (SOI) maintainability situated development onto which manageability arranged advancement practices and procedures can be mapped:

- Operational improvement (e.g., eco-productivity—consistency, effectiveness, doing likewise things better).
- Organizational change (e.g., new market openings—novel items,

administrations or plans of actions, doing great by doing new things).

- Systems building (e.g., societal change—novel items, administrations or plans of actions that are difficult to accomplish alone, doing good by doing new things with others).

4. **New concepts for sustainable value creation—negative forms of value:** Not many creators have contributed towards understanding the formation of new frameworks and producing an incentive over the value network in the sustainable business models writing by recognizing fizzled esteem trades towards understanding open doors for esteem creation. Yang et al., (2014) portray and characterize various types of significant worth (e.g., esteem nonappearance, value surplus, value destroyed, value missed). Rana et al., (2013) and Bocken et al., (2014) in their exploration propose a structure for plan of action development for maintainability by expressly considering worth pulverized and esteem missed inside the plan of action, as these frequently speak to critical open doors for manageability advancement. Their examination gives a subjective structure to encourage orderly investigation of the diverse types of significant worth for every partner.

- **Value caught**—current incentive;
- **Value demolished**—negative esteem results of current model;
- **Value missed**—esteem as of now wasted, lost or insufficiently caught by current model;
- **Value openings**—new open doors for extra esteem creation and catch through new exercises and connections (Figure 5.5).

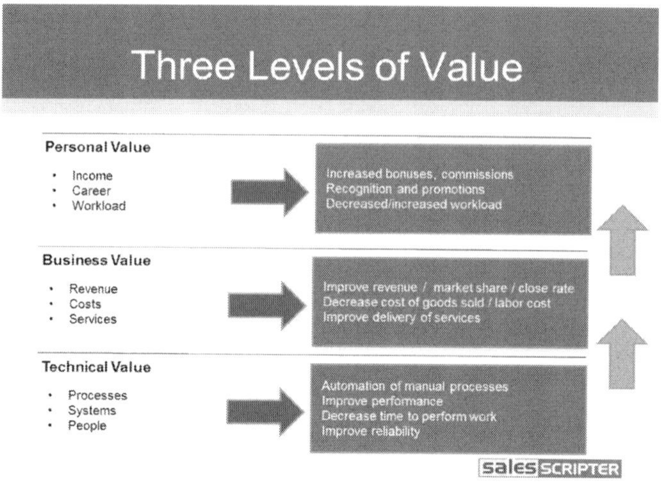

Figure 5.5: Three types of value propositions.

Source: https://salesscripter.com/3-types-of-value-propositions/

REFERENCES

1. *An Introduction to Sustainable New Product Development.* (2017). [eBook] (p. 13). Retrieved from: http://www.wrap.org.uk/sites/files/wrap/An_Introduction_To_Sustainable_New_Product_Development.pdf (Accessed on 29 March 2019).
2. Canciglieri Jr., O., & Teixeira, F. P., (2014). *Sustainable Product Design: The Development of a Conceptual Model.* Retrieved from: https://www.researchgate.net/publication/260105948_Sustainable_Product_Design_The_Development_of_a_Conceptual_Model (Accessed on 29 March 2019).
3. Clark, G., Kosoris, J., Nguyen, H. L., & Crul, M., (2009). *Design for Sustainability: Current Trends in Sustainable Product Design and Development* [eBook] p. 16. Retrieved from: http://file:///C:/Users/ADMIN/Downloads/sustainability–01–00409.pdf (Accessed on 29 March 2019).
4. Evans, S., Fernando, L., & Yang, M., (2017). *Sustainable Value Creation—From Concept Towards Implementation.* Retrieved from: https://www.researchgate.net/publication/312485712_Sustainable_Value_Creation-From_Concept_Towards_Implementation (Accessed on 29 March 2019).
5. Gökan, M., Marco, T., & Endris, K., (2011). *Assessment of Sustainable Practices in New Product Development* [eBook] (p. 11). Retrieved from: https://pdfs.semanticscholar.org/7543/e573de4eef2edbd2fe374d04d55fae7a24dc.pdf (Accessed on 29 March 2019).
6. Hallstedt, S., (2008). *A Foundation for Sustainable Product Development.* [eBook] (p. 234). Retrieved from: https://www.diva-portal.org/smash/get/diva2:833746/FULLTEXT01.pdf (Accessed on 29 March 2019).
7. http://www.sustainvalue.eu/publications/sustainvalue_slides_general_and_wps_090413.pdf. [eBook] (p. 24). Retrieved from: http://www.sustainvalue.eu/publications/sustainvalue_slides_general_and_wps_090413.pdf (Accessed on 29 March 2019).
8. Oertwig, N., Jochem, R., & Knothe, T., (2018). *Sustainability in Model-Based Planning and Control of Global Value Creation Networks* [eBook] (p. 8). Retrieved from: https://ac.els-cdn.com/S235197891730029X/1-s2.0-S235197891730029X-main.pdf?_

tid=fc9b8cee-d0c0–4173-b4cd-c8aa835e3fde&acdnat=1543837339_d99ff76cc03d11fc141c7ee9a9c51c90 (Accessed on 29 March 2019).

9. Pathak, P., Pratap, S. M., & Sharma, D., (2017). *Sustainable Manufacturing: An Innovation and Need for Future*. Retrieved from: https://www.researchgate.net/publication/317030927_SUSTAINABLE_MANUFACTURING_AN_INNOVATION_AND_NEED_FOR_FUTURE (Accessed on 29 March 2019).

CHAPTER 6

SUSTAINABLE MANUFACTURING FOR GLOBAL VALUE CREATION

CONTENTS

Sustainable value creation means the development of valuable products and services by using tangible and intangible assets while generating intellectual capital and ensuring transparency about environmental impact of the manufacturing process. There are five factors that can be called value creation levers. These are humans, process, equipment, organization, and product.

The value Creation Levers are integrated in the Value Creation Networks. The three dimensions of sustainability namely economic, social, and environmental are influenced by Value Creation Networks. Sustainable value creation can be pursued by innovative technologies that allow value creating levers to co-exist and affect others. These levers analyze value creation based on the three dimensions of sustainability, and hence they facilitate decision making.

The effect of these three dimensions on all stakeholders, in value creation through sustainable manufacturing (SM), has to be analyzed. There are various strategies that a manufacturer can adapt to facilitate to pursue manufacturing practices while creating value and these have been discussed in this chapter.

6.1. INTRODUCTION

Sustainable development (SD) can be defined as development that meets the need of the present, without compromising the ability of future generations to meet their own needs." Brundtland Commission came up with this concept of SD in 1987.

The business council came up with the publication 'Changing Course' for SD. This publication discussed why the business community should be concerned with sustainability and also how the businesses can participate in achieving this objective.

It was during this time that sustainability was recognized as an integral part of business continuity and crucial to strengthening the strategic position of the company. Since then there has a lot of discussion about ethical and compassionate ways of doing business with more accountability.

The concept of sustainability has three facets namely economic, environmental, and social. They can be called as the foundation of the concept of sustainability through sustainability can be interpreted in various ways.

Campbell stated that these three aspects gave rise to three conflicts:

- Grow the economy;
- Distribute the growth fairly; and
- In the process do not degrade the ecosystem.

Rockstrom suggested a new approach to achieve sustainability. He identified nine terrestrial boundaries, within which if a business functions, it would not cause any major man-made environmental impact on a large scale.

A transition towards sustainability is observed in manufacturing as a basic element of our society that involves complex systems and uses intensively materials and labor. As Miltenburg (2005) states "manufacturing is complex. Large numbers of employees – skilled and unskilled, line, and staff, flexible, and inflexible – work in a network of domestic and foreign facilities, formal, and informal systems, good, and bad practices, and old and new cultures."

Sustainability to a large extent is dependent on manufacturing practices. This is because manufacturing sector is dependent on resources and is responsible for its depletion. The toxic wastes and gases of the manufacturing units are released into the atmosphere, land, and water. Besides, IEA reported in 2008 that 33% of the energy consumption and 38% of carbon dioxide, both direct and indirect, can be attributed to manufacturing. Needless to say, manufacturing plays a huge role in global economy. For instance, manufacturing sector offers 70% of the job opportunities in Europe and contributes to20% of the continent's GDP.

Globalization of manufacturing sector has contributed immensely to the growth of economies. Almost 70% of the variances in income of over 100 countries can be explained by differences in manufactured product export data alone (World Economic Forum Report, 2012).

Therefore, the manufacturing sector can play a huge role to ensure that the goal of global sustainability can be achieved. Sustainability has to be embedded in manufacturing and can be perceived as "a systems approach for the creation and distribution (supply chain) of innovative products and services, that: minimizes resources (inputs such as materials, energy, water, and land); eliminates toxic substances; and produces zero waste that in effect reduces greenhouse gases, e.g., carbon intensity, across the entire life cycle of products and services" (Rachuri et al., 2011). The old paradigm assumes that resource is infinite and there is unlimited capacity for regeneration and forms the basis for the current manufacturing practices. In the context of sustainability, the manufacturing models have to be updated. Seliger et al. suggested three approaches that will make manufacturing sustainable. They are:

Technological innovations should pave the path for efficient use of scarce resources, thus ensuring manufacturing sustainability.

- The utilization ratio of a product and it inputs should be improved thereby improving the use-intensity of the products. Two approaches form a part of this strategy. First, Distributed use of products and its inputs. Second, service-oriented business models.
- The shelf life of manufactured goods can be extended by increasing the use phase or by the realization of multiple use phases.

Neely observed the global trends in manufacturing, by studying the time periods 2007, 2009, and 2011. He published his first study in 2008. It was based on data collected from 2007 and had detailed proofs about the scale of servitization. This percentage was 30 which implied that 30% of the businesses engaged in manufacturing also rendered services. The study also covered the 12 areas in which these businesses offered services.

It included design and development, leasing services, consulting services, financial services, outsourcing, and operating services, retail, and distribution services, transportation, and trucking services, installation, and implementation services, systems, and solutions, maintenance, and support services; procurement services; property and real estate and retail and distribution services. It also covered the extent to which various nations were servitized. For instance, almost, 60 percent of the U.S. manufacturers were servitized whereas in China the percent was less than 1 in 2007.

The third study which considered the 2011 data, the scale of servitization is still around 20 percent, and there has been no major change since 2007. The category of the services that manufacturing businesses rendered also remained the same the only difference being the volume of each service rendered.

The services rendered arranged in descending order by volume are as follows, systems, and solutions services, design, and development, maintenance, and support and retail and distribution services. Among all the services rendered, the manufacturing businesses offered these four services more as compared to the rest.

Diffusion of servitization by the different countries remained relatively unchanged in 2011 as compared to 2007, with the exception of China which rose to 19% from 1%. China has clearly displayed its intent to move up the value chain rather than focusing solely on manufacturing activities which added low value. Also, the servitization is U.S declined by a small percentage from 58% to 55%.

Turunen and Finne published a study recently that diffusion of servitization by manufacturers belonging to the same industry will be comparable because they operate in similar conditions. The differences in servitization between manufacturers vary because first, they operate in different industries; second, they operate in different geographical conditions and third, the environmental requirements differ.

The field of servitization still attracts academicians with different specializations as there are more areas that can be investigated. Lightfoot et al. enumerated five basic research concerns related to servitization,

- Service marketing, operations management communities, and service management are actively engaged in product-service differentiation.
- The operations management community has to prioritize competitive strategy as a main research concern.
- Operations management and service management mainly concentrate on creating customer value.
- The service science community is predominantly concerned with fostering customer relationships.
- The product and service customization are being addressed by all communities.

6.2. CONCEPT OF VALUE

The definition of the term value in Oxford Dictionary is "the regard that something is held to deserve; the importance, worth, or usefulness of something." The concept of value means different thing for different individuals.

Value has been defined differently by different authors pertaining to different fields. Slack defined value as "as a measurement of the worth of a specific product or service by a customer as a function of:

- Product usefulness in satisfying customer needs;

- Relative importance of the need being satisfied;
- Availability of the product when is needed;
- Cost of ownership to the customer."

He clearly asserted that the concept of value is subjective and emphasized more on the customer who buys the product. Among all the disciplines that the concept of value has been disserted in, business literature, in particular, has focused on it more.

Porter's value chain framework can provide a good basis for understanding and evaluating the logic behind firm-level value creation with respect to the profitability of a business. Many authors analyzed the logic behind implementing a plan in business by using value as the dominant concept.

Business theories have disserted about the concept of value a lot, and this led to the emergence of other allied concepts like value proposition, value creation, and value network and value chain. In connection with this, Wynstra, and Lindgreen (2005) recognized value analysis, value delivery and value creation as three major themes in industrial and business marketing.

The most important source of strategic advantage for a business can be derived from creating and delivering value. The value creation and delivery system describe this logic behind how this value creation happens. According to Ricart and Masanell (2010), there are two primary concerns in the concept of value in business literature, and it is linked the idea behind a business model. The issues are:

- How a business can identify value and satisfy a customer?
- How the business manages to make profits in the process of value creation?

The business literature has always identified consumers as the primary focus for creation of values. Holbrock identified eight categories of consumer benefits- esteem, excellence, play, efficiency, esteem, aesthetic, spirituality, morality, and politics. Woodruff, on the other hand, suggested that deconstruction of customer value can divided into product characteristics and attribute performance, customer objectives and purpose and consequence in use situations.

The definition of value as per Mason and Spring (2011) is the profit gained by a consumer from an exchange. Earlier only economic value interchanged between a consumer and the business was included in the scope of the term value. Now it has evolved to include in its ambit to the understanding of value as derived from goods and services but also from buyer-seller relationships (Lindgreen et al., 2012)

The term value includes not only transactions having a monetary impact but also interchange, associations, and communications. Some authors have also broadened the scope of the term value. This broader term includes the generation of value for the consumer, the business and the society.

Leapk et al., (2007) suggested a multidimensional outlook of the term value that states that value has different definitions at individual, business, and public levels. Also, value creation can be done by individuals, businesses, and public.

Den Ouden studied the all the aspects related to concept of value. The idea of Value Framework was thus conceived that considers value in different levels namely for the consumer, business, environment, and public. It also considers value from the standpoint of social sciences which includes the branches of ecology, sociology, economy, and psychology.

The company should take a holistic approach in pursuing sustainability and integrating it into the value creation lever. It should consider social and environmental impact and address the concerns of all the stakeholders including the society.

In the backdrop of global competition, producers, and customers have to come together to create value. This task cannot be solely carried on by the manufacturers of goods.

This broader view of companies' stakeholders and co-creators of value is embedded in the concept of value networks which expands from the traditional view of value chains and involves "complex dynamic exchanges between one or more enterprises, customers, suppliers, strategic partners, and the community."

6.2.1. Hart and Milstein

It is an intricate and multifaceted process to generate sustainable value. This was illustrated by Hart and Milstein in their proposed Sustainable Value Framework. These frameworks state different policies, factors, and the resultant benefits to the companies in two-dimensional axis.

Time is indicated by vertical axis, both short term and long term. The horizontal axis indicates the conflict a business faces between taking new approaches and new perspectives from external environment and safeguarding the internal efficiencies and technologies (Figure 6.1).

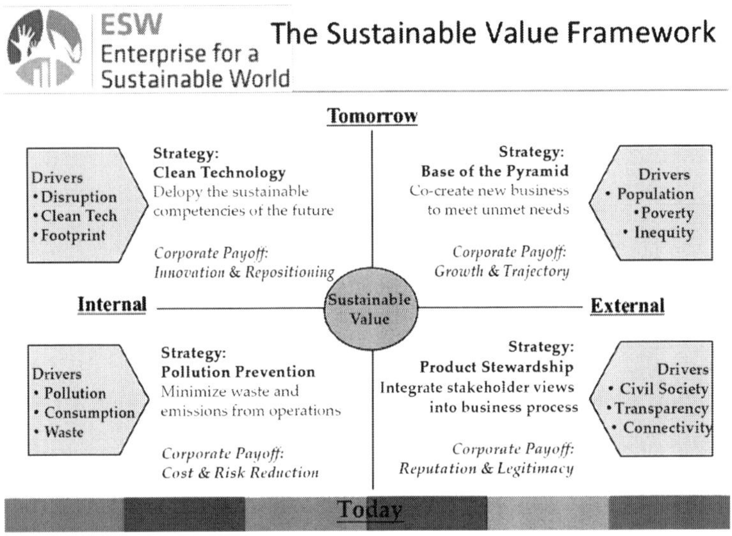

Figure 6.1: Sustainability framework.

https://www.google.co.in/url?sa=i&source=images&cd=&cad=rja&uact=8&ved=2ahUKEwi7hoPxtZnfAhVBQH0KHc1DCQEQjRx6BAgBEAU&url=https%3A%2F%2Fwww.slideshare.net%2Famitkapoor%2Fthe-sustainable-value-framework&psig=AOvVaw0nN5m0BN3XJwj7z7Fb9nQD&ust=1544674180249198

6.3. SUSTAINABILITY ENGINEERING

Sustainable engineering means to design and operate processes in a way that resources are used judiciously, and energy is consumed efficiently. The objective is not to compromise the ability of the coming generations to meet their requirements while ensuring that environmental impact is minimized.

Sustainable engineering is an emerging idea in the science of manufacturing. Globally manufacturing concerns should collaborate and incorporate sustainability as one of their strategic advantages. They should encourage innovation in technologies to ensure manufacturing processes evolve are able to maintain sustainability of our planet.

The emphasis is on transform the manufacturing processes and finds a solution to influence the human life positively.

According to the current research development, it has been realized that there is a systematic framework which cross the tracks in the way of sustainable technology and its assessment. An in-depth understanding of SM is required which would relate it with the valuation and mathematical modeling of manufacturing processes and equipments. It is also required that there should be virtual systems for product development. There has been little advancement for the organization of sustainable value creation in product and material cycles on different levels of aggregation.

These two standpoints are amalgamated for generating systems and procedures that will produce social capital. This will enable the people to self-learn and sustain themselves. The research carries various agendas, but the common objective is to find prospective areas to carry out initiatives that will help create awareness and about SD.

A lot of contributions have been made by various emerging organizations, but they remain unnoticed, and it is important to identify such organizations as this would help in promoting innovations which would further empower fair trade and competition. The implementation of sustainable value creation at a global level in mutual exchange of knowledge between partners from different communities should be specified or identified properly.

It is only the knowledge which cannot get diminished and could only get utilized and then expanded. A strong leverage can be expected from the manifold contents of knowledge management. As a result, services on information infrastructure or on public awareness for mutual exchange of ideas with societal stakeholders, and an integrated research training group for doctoral students from different continents becoming ambassadors of SM in their native countries and universities are possible.

With the help of modules and networks, it is possible to have full-fledged value creation which could be dependent upon the perspective of consideration. Modules in a bottom-up perspective constitute networks by dynamics of cooperation and competition within the different levels of aggregation from manufacturing cells via lines, factories, enterprises, local, regional, global consortia in value creation for tangible and intangible products.

On the other hand, in a top-down perspective networks integrate modules horizontally and vertically again on different levels of aggregation. E.g., original equipment manufacturers look for innovative suppliers of physical components or services thus improving their competitiveness.

6.4. VALUE CREATION FRAMEWORK

In the context of manufacturing sector, value is created by altering the input-output ratio. The inputs are the raw materials and other resources that go into the production.

The relationship between the consumer, producers, vendors, and other stakes in the business shape value creation at a global level. Also, the ability of these participants to spontaneously order themselves along with decentralized decision-making contributes to this value creation. Companies can create value by fulfilling the customer requirements of delivering goods or rendering services and the needs of those who have invested in the business. Other stakeholders like the vendors, manufacturers, and retailers are also benefitted in the process of value creation.

When stakeholders are integrated horizontally, materials flow through a supply chain. When manufacturing processes are integrated horizontally, inputs transform into outputs along the various stages of production. The major activities in the lifecycle of a product are procuring raw materials, transporting them to the production site, processing the materials, assembling the parts, producing the finished product and finally put it to its intended use. It also includes a host of other post-production activities recycle disassembly and remanufacture.

There are five factors that influence value creation namely product, process, equipment, organization, and human. The network of value creation includes modules of value creation that has been integrated horizontally and vertically. Value creation modules are customized as per the scope involved in specific case, representing the various phases of a life-cycle.

6.5. ASSESSMENT OF VALUE CREATION

Manufacturing needs an intricate system, and it comprises of designing, regulating, and supervising the value chain. Performance measurement acts as an important feedback that facilitates the organization to improve the process and ensure strategic advantage. Conventionally, the concept of value creation networks emerged from the dissertations of manufacturing systems. The factors that have driven this are the need to achieve consumer gratification, decrease the cost, and enhance the quality, increase efficiency in a time-bound manner.

Indicators and metrics are important because they express or measure the attributes of a product or process in quantitative terms. This facilitates comparison and benchmark settings. They simplify data which is complex and present them for evaluation. The challenge is the unavailability of a standard metric system. Each case has characteristics that depend on a specific perspective, which determines the types of measures that are relevant for the assessment. The measures must be sufficient and representative so that it can incorporate sustainable commitments.

6.5.1. Model-Based Approach for Assessing Value Creation

An organization strives to devise and adopt procedures that can generate more revenues while pursuing sustainability. Profit generation is as important as achieving sustainability and both influence each other. The traditional manufacturing processes have to be transformed, and this implies that greater benefits will accrue to contracted stakeholders.

If value created is extended to the societal shareholders, sustainability improves. Model-based approach includes PDCA cycle. It compares real or proposed options in terms of strategic alignment and sustainability performances to provide support for short-term decisions regarding single or multiple VCFs for a given project.

The approach tries to express the company's requirements into a representative set of characteristics, which should portray the relationship between different value creation levers.

6.5.2. Framework for Case-Based Application

The model-based approach should be systematically put to practical application in the following steps on a case to case basis. This method determines whether a value chain is sustainable and performs adequately. The following three steps are repeated till the objectives are met reasonably:

Step 1: Outlining the scope: The value created by existing methods is analyzed to determine the approach that will be adopted. Specific issues are addressed like opportunities, threats, weakness, and challenges. For each case, the stage of production and value creation levers are also recognized.

The value creation levers and their hierarchy molds and guides a value creation module. If the organization is not satisfied by the quantum of value created, the related value creation lever has to be modified so that the requirements are met. Attributes are properties that point out the present level of progress or the rate of change in development towards achieving the goal set by the organization. The set of attributes with different priorities determine the amount of change in the predetermined VCF.

Step 2: Evaluation of the decision: The value creation levers and each phase of the lifecycle are represented by characteristics. These characteristics should also have the required metrics and sub-characteristics that can enumerate the value of the factors. An evaluation of change in one characteristic affects both the sustainability and the strategic positioning thereby also the related value creation levers.

Step 3: Choosing the best alternative: The established is used to evaluate the current situation. The current or planned options are formulated to overcome the barriers at the final value creation level and assessed to check if objective is achieved. If not, these steps are repeated as required based on the results and their effect on value creation. The best option that would meet the objectives is identified and applied.

6.6. SUSTAINABLE VALUE FROM INNOVATIVE PRODUCTS, PROCESSES, AND SYSTEMS

SM can be achieved by bringing innovations in products, processes, and organizations. The manufacturing processes have to shift from conventionality and adopt a more holistic approach to the practices. The quality of a product must be evaluated separately autonomously from the performance qualities and traits.

Simultaneous engineering methods are designing methods for products that run concurrently and not sequentially. They form the basis for combining the product design process with the process design process. Nonetheless, if a team of experts from various fields and countries are included in the process, they have to be managed centrally, failing which, the experts will find it difficult to transcend the boundaries of their organizations and contribute meaningfully towards developing innovative products, procedures, and structures.

It is imperative for all the stakeholders to cooperate and work collectively on the mutual goals with complete dedication, so that the manufacturing process can be approached holistically. SM has to be fully integrated, which means that all the sub-processes should function together as one process. Integrated manufacturing will facilitate the production of sustainable products and ensure that the processes involved do not have environmental impact.

It is expected that innovations will facilitate the development of an integrated value system that will support SM along with other value propositions like socio-economic value, socio-environmental value, technical value, and socio-political value, etc. that the integrated system will contribute.

6.7. ROLE OF SUSTAINABILITY IN VALUE CREATION

Various studies have proved that efforts to minimize environmental impact or sustainability can have a positive impact on the economic performance of manufacturing company. It is evident in the overview of various related studies that the relationship between environment and economy is labeled differently.

The different labels used can be premium profits, company's performance, strategic advantage, sustainable value creation, sustainable economic value creation, economic performance.

The business models for sustainability often differs; however, their actions are comparable and have the same interfaces. The different labels mentioned above support this claim. Though the actions taken for achieving benefits of sustainability may differ, nonetheless the financial outcome is not different. This outcome is either generating revenues or cost cuttings obtained, both of which clearly are associated with value creation.

Another perspective of value is premium profits. Premium profits of a company can be attributed to specific unique recourses which help in generating the additional profit. This resource-based dimension is an add-on the notion of traditional profitability.

The concept of premium profits highlights the benefits that an organization derives by focusing on the environmental performance since it has an impact on the economic performance of an organization. Some academics believe that premium profits are actually rents that an organization which owns a scarce resource derives and these are specific to that organization, unlike economic profits that accrues from strategic positioning.

We can compare the contrasting effects of generation of premium profits with accruing of rents and in doing so also compare the rents with levers to premium profits. Premium profits can appropriately determine value in the context of the role of sustainability in value creation in manufacturing business. It can be argued that revenue increases and margin improve if an organization adopts sustainable practices and this is also a source of strategic advantage to the organization.

These are used for measuring value creation to make sustainability look like an attractive proposition for ensuring economic benefits. The drivers for value creation are risk management, pricing power; human resource management, cost reductions, and these drivers also ensure margin improvements. Market share and access to new markets can leverage a company's position and generate more revenue.

6.7.1. Resource-Based View of Value Creation

A resource denotes an asset or raw material owned by an organization, controlled by it, or that can be accessed by it and which goes into production. Resources may be physical, intangible or human resources.

The resource-based orientation leans emphasize on the company's ability to exploit its specific resources and assets and also how it is managed. The way the resource is managed and exploited by the company directly impacts its ability to create value in a sustainable manner.

Experimental studies that take a resource-based view of value creation believe that manufacturing approaches that consider environmental impact and capabilities that have a strategic advantage are linked. This is because the organization develops their unique resources. A company's tactics for staying competitive and economic performance is linked to the unique resources in its control and other proficiencies.

Resources are diverse with respect to a company and include assets like technology, firm attributes, processes adopted and casual relations.

Valuable resources that can help a business explore new opportunities and negate the adverse conditions in its environment forms the foundation of a company. Usually, there are no comparable substitutes for these resources, and they are irreplaceable.

Resource is a phenomenon that has its unique properties, which may be an asset or raw material, and must be exploited and controlled by the competences of the organization so that premium profits can be earned. The competencies of an organization lie in its organizational capability. As per Helfat & Peteraf definition "An organizational capability refers to the ability of an organization to perform a coordinated set of tasks, utilizing organizational resources, for the purpose of achieving a particular end result."

The knowledge about resources and organizational capability leads to the belief that an organizations strategic position is strengthened if it adopts a unique strategy for value creation which gives it a competitive advantage over its rivals. As discussed earlier, the economic performance and environmental performance of the resources are related, and hence the organization's resources and capabilities can provide various strategic benefits.

Thus, competitive advantage can be considered an outcome of unique resources and capabilities of a business, and this advantage positively impacts the capacity to earn premium profits. Each organization exploits its resources in its own way, and hence premium profits vary between companies adopting a similar approach to sustainability.

In the context of sustainability, resources generate different strategic advantages depending on the volatility of the business environment. Conventionally, resource-based views did not include environmental impact. This view is addressed by Hart (1995) which proposes the notion ‗the natural resource-based view of the firm.'

According to Hart, ecology can be viewed as a source of strategic advantage. He is of the opinion that environmental limitations should be amalgamated into the company's management process. This will generate new resources and develop proficiencies which will add a strategic benefit to the organization. The concept of sustainable value creation includes the concept of environmental opportunities and limitations.

Environmental factors should always be considered before assessing or developing resources and capabilities within an organization if the goal of sustainability has to be achieved. This approach has been termed as dynamic capabilities and emphasizes on the importance of exploiting the firm-specific resources both internal and external to cope with the changes and make use of opportunities in the environment. There are some factors that are contingent, and they play a vital role in recognizing and utilizing the company-specific resource and proficiencies.

6.8. REVENUE GROWTH AND SUSTAINABILITY

Sustainability has a positive impact on revenue generation. Environmental issues worldwide force companies to take steps to either negate the effects of the damage they cause or bring innovation in technologies to limit the damage. Compensatory measures include reducing emissions and acquisition of credits. Innovative technology could be used to improve processes, develop new products, and try out new product combinations.

A business should focus on innovation to survive in a hostile environment and also find out new channels of revenue generation. Boston Consulting Group opined that a company should first try to operate legitimately within acceptable social norms and take appropriate actions; only then it can pursue sustainability goals.

The pragmatic study conducted by Dangelico and Pujari proposes that the prospects of a business are not always restricted due to external regulations. Regulations are important because they offer avenues for sustained revenue generation and prospects for new opportunities. Hence it is important that stakeholders view sustainability as a facilitator of earning profits.

Businesses should feel encouraged to attain sustainability because this way it can follow market growth and thereby accelerate its own growth.

When an organization merges sustainability into business processes, it can expect certain financial benefits related to increase in revenue, to accrue to the firm. This benefit could be in the form of increased turnover, access to new markets, product differentiation, higher brand value, better return on investment and better strategic position.

Increase in the turnover can be achieved in two ways: first, by increasing the market share for the existing products or introducing new ones in an existing market, second, by accessing new markets and delivering products there. These two strategies of increasing market share and entering new markets can be termed as the value creation levers.

6.8.1. Market Share

Resources that are unique provide competitive advantage to the organization that owns it. Such organizations enjoy bigger market shares and earn more profits. A resource would be considered unique if the competitors do not have access to similar resources or cannot duplicate it at justifiable costs. Resource profiles and market shares are seen as contingent in the sense that if resources change the relative market shares typically changes as well.

It is vital that organizations either recognize threats from rivals that are going beyond compliance with R&D capabilities or have an important resource such as their image of being a sustainable firm. It is also important for the organization to acknowledge that the relative market share is dependent on some resource profiles.

Image of an organization is one of the determinants of the profit earning capacity of an organization. An organization which has the reputation of being environment-friendly will be able to attract consumers who value environmental issues. A reputation for abiding by environmental policy can provide competitive advantages. Hence it is important to not just develop these policies but also adhere to them.

Large hotel chains like Marriot, MGM acknowledge the role of their responsible corporate behavior towards environment is partly the reason behind earning huge profits. This is a good example to demonstrate the interdependence of profit and corporate image. An effective strategy for achieving sustainability is vital for increasing market shares.

An organization has to work hard to earn the reputation, but once achieved, the reputation is a unique, valuable resource and also a source for competitive advantage. There is no uncertainty about the fact that there is a growing concern for environmental issues; similarly, ample market opportunities are also available. Consumers are more conscious these days, and they tend to prefer green alternatives to conventionally produced goods.

Organization is under pressure to lean towards green manufacturing in so that they can ensure consumer loyalty and also attract new ones. Greenmarket is emerging fast and will get bigger in the coming days. This transformation in the society and changes in expectation make innovation of environment-friendly products inevitable.

It is evident from market data that the demand for ethical goods is on the rise and has remained so for some time now. The market for ethical products was estimated at $36 billion in 2009. The expenditure on these goods has also multiplied in the last decade.

6.8.2. New Market Entry

Newmarket entry is the ability of an organization to penetrate new markets and thus find new sources for revenue generation. A product has to be redefined in order to cater to new market segments, innovate new products that can create new segments in the current markets. This is called research and development.

When an organization integrates environmental limitations into its processes, it generates various efficiencies and resources like human resource, technology, and other intangible assets.

R&D offers immense opportunities to an organization and increases its turnover. This can be attributed to the three opportunities R&D presents:

- Incremental R&D facilitates into an existing market.

- Radical R&D enables admission into or development of a new green segment of an existing market.
- Isolated R&D helps in accessing a completely new market.

Organizations that adopt sustainability can expect increase in profitability. This is because sustainability has a positive impact on value creation levers. Thus, the financial case for sustainability leading to revenue growth can be created with a focus on how to exploit, develop, and innovate on current and new capabilities that manage company resources as well as exploit, develop, and innovate on the resources themselves.

In order to enjoy premium profits by integrating sustainability into manufacturing process, an organization has to analyze the existing resources against an external ecological context. It also has to acquire and build new resources with which it can manage the ecological context. Organizations, while in the process of managing resources also develop new efficiencies. This results in spill-over and fresh prospects for increasing market share or gaining access to new markets.

It is evident that when new resources or capabilities are integrated into the process, value creating factors of the organization derives a boost from it. Increased productivity and capabilities are not the only ways an organization is benefitted by adopting sustainability. The image of the organization also enhances, and company is also perceived as being more transparent in its operations. Consequently, many benefits accrue to the organization like lower capital cost, reduced labor cost, avoidance of external threats, green taxes.

As such, in the case for revenue increasing value levers, the actions within the market share increasing value levers affect the possibility for new market entry positively due to developed capability bases. Strategies that can improve profits follow actions aimed at increasing revenue and vice versa.

Steps taken to improve efficiencies in processes would ensure rise in margins and thereby increase in premium profits. It also brings about incremental changes in existing line of goods which could lead to market share gains or access to new markets. On the other hand, when an organization adopts sustainability, it enjoys reputational concessions like reduced labor cost, lower direct costs.

Hence, sustainability brings numerous benefits to an organization. The correlation between value levers serves as a knowledge base for a transformation in the perception of sustainability.

Sustainability has an impact on all value creation levers of an organization at different levels. When an organization acknowledges this fact, it is motivated to adopt green practices.

6.9. SUSTAINABLE MANUFACTURING PATTERNS

Manufacturing can be defined as a process by which goods are produced, and services are rendered, which an integral for maintain the standard of living of humans and advancement of economy. The term manufacturing has a broader connotation than the term production because it includes all types of services as well.

The growth of an economy of any nation depends on its manufacturing capacity, the types of goods available and its export capacity. Hence manufacturing is vital for wealth creation and the well-being of the nation's population.

The role of manufacturing sector in creating job opportunities, societal wellbeing and developing economy in developed as well as emerging economies is crucial. It cannot be denied that manufacturing stimulates economic growth; however, it is also important to consider the impact of manufacturing on the ecosystem. Manufacturing requires high amount of energy consumes lot of resources, emits harmful gases and other toxic wastes. This necessitates organizations to take a sustainable approach towards manufacturing.

Manufacturing units have to adopt design for environment (DFE) so that it produces less waste and emissions. The aim is to adopt an approach for product development that integrates the environmental requirement throughout the lifecycle of the product.

The entire manufacturing process has to undergo transformation in order to attain sustainability objectives. Analyzing only the product designs and manufacturing routes will not suffice.

The ecosystem, society, and economy are the three facets of sustainability. It is also referred to as triple bottom line (TBL). These three facets are crucial for addressing sustainability in manufacturing process. Another well-recognized concept in sustainability is the three pillars- profits, planet, and people.

Technology can be considered as a compassionate dimension in SM. Technology is an integral element of SM model. Technology can bring about a positive transformation in the social, financial, and ecological components of SM. The major participants in SM provide the required stability to the process. Education and technology contribute positively to sustainability and are crucial for gaining financial, ecological, and societal benefits.

The role of education is to generate sustainable value. There is an urgent requirement, in countries worldwide, to reorient educational needs. There is need to spread knowledge skills and technical know-how that can support sustainable consumption paradigms in SM.

Ethical practices and answerability are crucial facilitators for attaining sustainability and environmental growth. A new model for SM is required where qualified experts can use technology and accountability for discovering new methods of coping with sustainability issues.

Morals guide the manufacturers to adhere to ethical norms and accountability confirms the answerability, of manufacturers to the environmental and social issues (Figure 6.2).

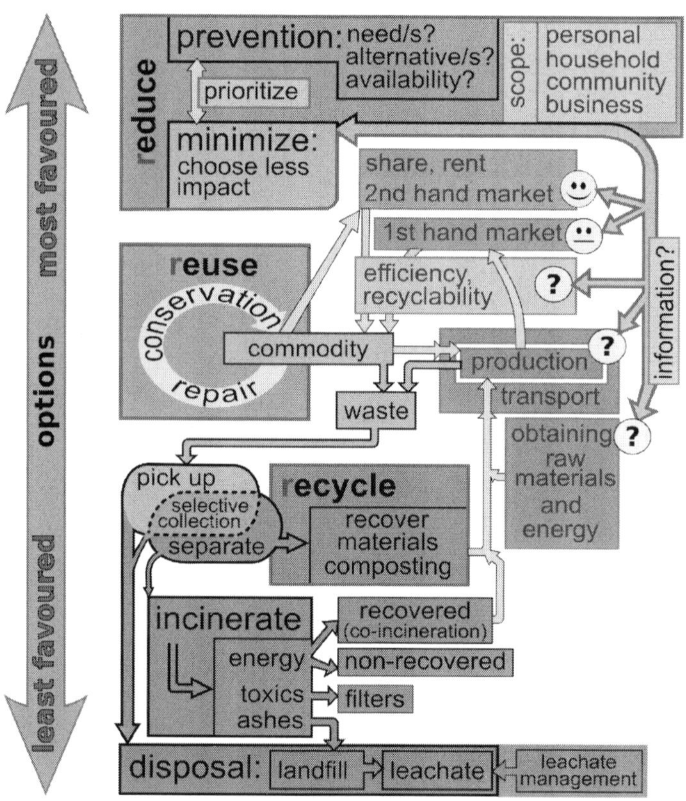

Figure 6.2: Waste minimization.

Source: https://en.wikipedia.org/wiki/Waste_minimisation#/media/File:Waste_hierarchy_rect-en.svg

6.10. STRATEGIES TO ENSURE VALUE CREATION THROUGH SUSTAINABLE MANUFACTURING (SM)

Manufacturing generates value through the actions and exchanges between producers, vendors, consumers, and other stakeholders. It is important to analyze the financial, ecological, and societal impact on all stakeholders in value creation through SM. This concept is also called the TBL.

Lazlo opines that sustainable value is created when value is created for all the stakeholders concurrently. Stakeholders include the shareholders, manufacturers, vendors, consumers, and other participants in the manufacturing process. The TBL standpoints and the desires of the stakeholders could be incompatible sometimes. Therefore, it is important that the manufacturing company balances the trade-offs between these two conflicts when it is on the path of SD.

There are various barriers that the company has to overcome and develop advanced tactics to ensure that it does not deviate from its goal of SD. There are various strategies that a manufacturer can adapt to facilitate the integration of SM practices in the production designs some of which are discussed below.

6.10.1. Role of Education and Workforce Development

SM policies can aid in improving TBL value for all interested parties. However, it often happens that the scope of SM is narrowed to comprise only of the ecological impacts and reduction in consumption of energy in the manufacturing activities.

One of the chief causes of this is that sustainable practices are not included in the curriculum. Consequently, the syllabus across schools, colleges, and other educational programs do not give priority to the SM practices.

Corporate knowledge and innovation are the two pillars of SM. A workforce that has been imparted sustainable education will be able to contribute meaningfully in these two areas. Sustainability education and workforce development must concentrate on the 3Rs of the education lifecycle of personnel, namely- Recruit, Reeducate, and Restrain. This will empower the workforce to innovate and develop new technologies and also improve their knowledge and hone their skills.

This life-cycle based approach in education and skills development will ensure that there is an uninterrupted supply of capable workforce which is also well trained in sustainability concepts.

6.10.2. Tools to Support Decisions of the Future Generation

Lifecycle Assessment is often perceived as a peripheral activity to analyze the potential impacts on the environment. It is not considered in unified way with the product and process development to give assistance for making decisions.

SM and Lifecycle assessment should not be viewed as an isolated activity and should be integrated into the tools for analyzing decisions. This would make SM a component of value creation and resource optimization.

The challenge for industry is that many decision support toolsets are contained in disparate systems and they lack interoperability. Lifecycle data is not always readily available. It is important to make decision support tools cohesive so that entire lifecycle sustainability can be tested at an early stage in the development of a product. This will assist in analyzing the feasibility of the product.

Use of interoperable decision support tools provides information of the entire lifecycle of a product and facilitates innovation at the product, process, and system levels. This also enables the appraisal of the sustainability impacts of the total lifecycle and assists in taking appropriate decision. What-if analysis can be used to study the impact of changes caused by the use of various inputs, processes, and supply chain policies on the environment, economy, and society.

6.10.3. Risks, Uncertainty Involved in Supply Networks

Most organization faces at least one supply chain disruption every year. Consequently, it has to bear loses, production is stalled, and at times even the reputation is damaged. Supply chain may internal such as damage to important components of an equipment or external such as vendors going on strikes.

Such events may also impair the long term efforts to achieve sustainability and cause instantaneous loses. The supply network is becoming increasingly complex, and hence there is always a risk It is crucial for the organization to equip themselves with capability to foresee and avoid to loses if possible or develop the capacity to mitigate the loses.

The ability to establish safe operating envelopes, well within the bounds of acceptable performance, and to operate within those boundaries, assures safe, energy-efficient, environmentally-responsible, and sustainable operation. The risk evaluation frameworks that are being used currently do not touch upon all factors or its aggregations thereof.

The current frameworks also lack depth in perception, are crude and analyze risks based on past happenings. Therefore, risk models should be able to observe and analyze events precisely when they occur and should be able to identify, assess, evade, and mitigate risks. This approach is necessary to create value through sustainability and also increase the capacity to tackle the uncertainties and disruptions.

An organization should be able to predict threats of disruption in supply chain and implement strategies that can mitigate those risks. This will ensure that the organization is able to satisfy the consumers and also generate revenues while pursuing sustainability.

6.10.4. Increasing Product Lifecycle Management (PLM) Capability

The conventional practices for planning process only focus on the manufacturing phase and not the entire lifecycle of the product. Process planning should also include the assessment of product design and its impact on the other phases of the product lifecycle. Integration Product Lifecycle Assessment into the planning process will help in value creation.

The perception of process planning as a tool, to produce products with specific characteristics needs to change. The process should be efficient and merged with the designing phase of the product to ensure end-of-life (EOL) processing. The process planning should also encompass recycling, remanufacturing, and reuse.

A holistic approach towards the manufacturing process is needed, and all the possible methods should be considered. The planning process should include sustainability goals for all 6R processes. The 6Rs include rethink, reduce, reuse, reuse, recycle, replace. The challenge in exploring all the alternatives is, classification of all the inputs and process is required to access all the data is not always possible. Planning processes for the entire lifecycle requires a lot of knowledge. Generating such advanced product lifecycle management (PLM) capability will increase the sustainable value creation and reduce the requirement for building a complex process.

6.10.5. Lifecycle Cost Models

A good lifecycle cost model would predict the costs and the capability of the product and process to create value more precisely. It can also help in reduction of costs. It is crucial that all the determinants of product costs are analyzed early in the lifecycle of the product so that all the possible alternatives can be evaluated, and a decision can be taken regarding the same.

The cost of a product can be estimated by evaluating past data. This approach is restricting in nature and does not encourage innovation. The correct approach is to thoroughly understand the basic cost elements which will facilitate the development of precise cost models that can be used to analyze all product alternatives.

A good cost model should include information from all the phases of product lifecycle. The cost model will enable precise cost evaluation in the initial stages of the development process and help in saving costs. It can also facilitate innovative product designs that make the end-life management of the product more efficient. When a product can be reused, recycled or remanufactured the salvage value of the product increases.

The total lifecycle approach has added benefits like reduction of unforeseen escalations in input costs. This becomes more pertinent where the manufacturer is also responsible for treating or disposing the products after use. This is called extended producer responsibility.

6.10.6. Flexible and Scalable Manufacturing Alternatives

Manufacturing processes have undergone transformation in the last decade, and the supply chain function now work in a cohesive manner. The manufacturing companies should now embrace flexible and scalable systems. This will enable the setting up of production facility at the most suitable location where resources can be accessed easily, and best equipment are available. The company would enjoy reduction in costs, increase in productivity and benefits of sustainability.

The current processes and structure will sustain flexible and scalable manufacturing. This is because they cannot adapt and aid in pursuing sustainability. A Reconfigurable Manufacturing System (RMS) can be used to cope with the rapid changes in the structure of the company. RMS can quickly adjust production capacity as per changes in requirements. RMS can facilitate 6R based closed-loop material flow and hence help the organization in risk minimization and customer satisfaction.

Flexible and scalable manufacturing alternatives can also reduce transportation cost and the sustainability issues associated with it. These strategies are good corporate practices since it will enable reduction of costs and improve margins simultaneously enforcing sustainability. A standard has to be developed that can gauge the TBL achievement and facilitate comparison.

The provision of such methods would enable the common-scale and consistent evaluation of performance; Standard metrics will make the products and processes comparable, in terms of sustainability achieved, across different countries culture and trade practices.

It will act a yardstick to measure sustainability in order to determine rewards which in turn will further motivate the organizations to improve product and process sustainability and reduce the lifecycle costs of the products and enable its reuse/recycle. Europe is the first in this field of working towards establishing a standard metric for measuring sustainability. Global Reporting Initiative is one such example by which Europe is attempting standardization.

Standard metrics and measurement frameworks will facilitate in accurate evaluation of the outcome of sustainability-related initiatives. It will also help to focus on areas that need improvement. It supports in decision making by assessing the long term and short term benefits of pursuing sustainability. This way metrics help strengthening the strategic position of the organization by creating sustainable value.

Recently efforts have been taken to form an inclusive model and standard metrics for evaluating the performance of SM throughout the product lifecycle. These methods also enable companies to evaluate the outcome of alternative strategies. Comparability is achieved, and it helps in better evaluation of possible outcomes. The ability to evaluate performance along the core aspects will enable better decision making considering TBL performance.

6.10.7. Other Strategies

Besides the strategies mentioned above, there are other areas where strategic capability can be enhanced to ensure value creation by adopting SM practices.

In current systems, knowledge capture is often not systematized, nor does it cover the entire product realization spectrum of the total lifecycle, necessary for intelligent and sustainable design and manufacturing.

Decision making based on real-time data will make the decision-making process more efficient and also enable to save costs. The company can foresee unintended consequences and alleviate uncertainties. Better foresight and ready access to information will make the operation more efficient by removing redundant processes. It will also help the organization to pursue sustainability and enhance the reputation of the business.

Cyber-attacks are affecting nations, and it poses the risk of completely disrupting the supply networks. This is another area of concern that needs strategies so that SM can be pursued.

It is crucial that nations safeguard themselves from these attacks. Similarly, it is also important that the supply chain is not exposed to risks beyond the acceptable level and is able to function properly. The policymakers have to create a secure environment and provide impetus to deal with these external threats.

Protected collaboration platforms are required that will enable secure exchange of information across supply networks. These challenges cannot be dealt with at individually by the organizations. Public-private collaboration is required, and the organizations should work towards achieving a common goal and creating value.

The industry required funds to conduct researches to guarantee industrial value of the deliverables. The public-private partnership can enable this. For instance, the U.S. recently established the National Network for Manufacturing Innovation (NNMI). These institutes are a product of public-private collaboration and are established with the objective of generating high-end technologies that can transform the manufacturing sector.

These technologies require funds and cannot be developed by any organization independently. U.S. has established numerous NNMI institutes that have developed various technologies. The objective now on is to establish NNMIs that will generate new technologies that support SM by reducing input requirement, making processes energy-efficient and minimize environmental impact.

If corporate assets are managed as a coherent unit, it will enable better decision making, make operations stable and ensure sustainability of the organization. The various elements that comprise the corporate assets are represented by functional departments in organizations, and there is seldom an awareness of the need for a systems approach to asset management.

A holistic approach to asset management will ensure that resources are readily available as per requirement, will help reducing costs through reliability engineering, reduce the adverse effects of risks, safeguard against interruptions, build a brand value and strengthen the business position.

REFERENCES

1. Badurdeen, F., & Jawahir, I., (2017). *Strategies for Value Creation Through Sustainable Manufacturing*. [eBook] p. 10. Available at: https://uknowledge.uky.edu/cgi/viewcontent.cgi?article=1003&context=ism_facpub (Accessed on 29 March 2019).
2. Bilge, P., Badurdeen, F., Seliger, G., & Jawahir, I., (2014). Model-based approach for assessing value creation to enhance sustainability in manufacturing. *Procedia CIRP*, [online] *17*, pp. 106–111. Available at: https://www.sciencedirect.com/science/article/pii/S2212827114002972 (Accessed on 29 March 2019).
3. Granados, M., (2014). *Sustainable Value Creation in Manufacturing Through Maintenance Services*. [eBook] p. 209. Available at: https://www.politesi.polimi.it/bitstream/10589/98504/1/PHD%20THESIS%20-%20Holgado%20Granados%20-%20October%202014.pdf (Accessed on 29 March 2019).
4. Jawahir, I., Badurdeen, F., & Rouch, K., (2013). *Innovation in Sustainable Manufacturing Education*. [eBook] p. 8. Available at: https://www.gcsm.eu/Papers/28/0.3_Jawahir.pdf (Accessed on 29 March 2019).
5. Moldavska, A., & Welo, T., (2017). The concept of sustainable manufacturing and its definitions: A content-analysis based literature review. *Journal of Cleaner Production*, [online] *166*, pp. 744–755. Available at: https://brage.bibsys.no/xmlui/bitstream/handle/11250/2465987/1-s2.0-S0959652617317213-main.pdf?sequence=4 (Accessed on 29 March 2019).
6. *Sustainable Manufacturing Transformation*, (2018). [eBook] Navigant Consulting, Inc., p.6. Available at: https://www.navigant.com/-/media/www/site/downloads/energy/2018/navigant-sustainable-manufacturing-transformation.pdf (Accessed on 29 March 2019).
7. *Sustainable Value Creation in Manufacturing Networks*, (2018). [eBook] p. 24. Available at: http://www.sustainvalue.eu/publications/sustainvalue_slides_general_and_wps_090413.pdf (Accessed on 29 March 2019).

CHAPTER 7

ADVANCES IN TECHNOLOGY FOR SUSTAINABLE MANUFACTURING

CONTENTS

A manufacturing business has to focus on the trends in the industry to have strategic advantage in the global market. 'Green Technology' is a term that is reverberating in all industrial sectors. The convention production processes have been emitting greenhouse gases in the last 100 years. Global warming- extremely detrimental phenomena for the environment- can be attributed to the apathy towards the ecological system. The toxic wastes that are drained into the water bodies, dumped on land or released into the atmosphere have also caused irreversible damage to this planet. Hence, countries all over the world have mandated many sustainable manufacturing (SM) practices and environmentally benign operations in every industry. Manufacturing facilities using conventional methods consume a lot of energy. Energy consumption is another important aspect of SM. The objective of innovations in technology is to rationalize consumption of scarce resource, reduce emission of toxic gases, minimize wastage, encourage reuse and recycling of materials. This will, in turn, reduce the cost of productions and increase profits. This chapter discusses the advancement in technologies with respect to green manufacturing, the importance of innovation in technology and the trends and future prospects of technology.

7.1. INTRODUCTION

Manufacturing sector is one of the primary sectors which have to adopt SM practices on a priority basis. Conventional manufacturing practices are paving the way for green technologies that will reduce the environmental impact and benefit Earth in the long run.

Plunkett Research defines green technology as the "application of advanced systems and services to wide variety of industry sectors to improve sustainability and efficiency." It is expected that green technologies will make the manufacturing processes more energy efficient, conserve scarce resources, reduce emission of greenhouse gases and other toxic wastes and also promote recycling of materials and encourage reuse.

There Green technology can be divided into various categories. Energy technology encourages use of alternative sources of energy and makes manufacturing processes energy efficient. Water technology aims to recycle waters used in industrial process. It also minimizes the consumption of water. Technological innovations are aimed at making alternate sources of water usable, like desalination of seawater. Pollution control devices have changed the way waste is managed. They emphasize on reusing and recycling wastes from manufacturing process. They also control emissions and negate the toxic wastes that are released into the atmosphere. Innovations in technology are create new product, process, and building designs in order the increase efficiencies in heating, air conditioning, and ventilation products. Also, technology has to meet the regulations that are formulated governments of various countries to support SM.

Skills and expertise are required to make the production of green technology cost-effective and efficient. There are various factors to be considered. First, the number of components of the technology should be few. Second, the technology should be efficient in mass production facilities. Third, the technologies have to align themselves to industry standards.

These technologies intend to reduce the cost to manufacture green technology to the point where it is a sustainable business in itself. The demand for Green technologies is growing, and more industries are shifting towards them.

The face of conventional manufacturing is transforming to integrate green technologies that besides having environmental benefits are also more efficient and have cost reduction benefits (Figure 7.1).

Figure 7.1: Adoption of green technology.

Source: https://pixabay.com/en/green-environmental-house-abstract-1357925/

7.2. TRENDS

Technological innovation is thriving, and the focus is more on producing sustainable solutions. The World Economic Forum has listed the top ten promising technologies of which many are those technologies that intend to minimize ecological impact. Some of the technologies that have made it to the list are water purifiers that consume less energy, nanoscale engineering for accurate drug delivery and heightened nutrition for health benefits at a molecular level.

As per Global Green R&D report published in 2012, the period 2007–2012 witnessed investment in clean technology and eco-friendly solutions by private players to the tune of $3.6 trillion. Of this total investment, $2 trillion was dedicated to renewable energy, $241 billion was invested in the R&D for green technologies, and $700 was for green construction, $231 billion for technologies that are energy-efficient and $ 238 billion for renewable sources of energy.

The market for renewable energy was projected at $ 248 billion in 2013. As per the Clean Energy Trends Report published in 2014, by 2023, this market is expected to reach $ 398 billion. The largest market is for biofuels (at $98 billion), followed by solar and wind. (at $91 billion & $58 billion respectively). In the year 2013, globally the manufacturers installed more new solar photovoltaic generators compared to wind energy generators.

This rapid growth is being fueled by significant investment in research and development and breakthroughs in sustainable technologies, as indicated by a spike in patent applications.

In the last five years, there has been a surge in the number patents related to climate technologies as per the World Intellectual Property Organization. The average global rate had a growth rate of 6% between 2006 and 2011. The climate technologies progressed at the rate of 24% during these years.

Developing economies have also displayed sustainable technological solutions. In the field of solar energy, top 20 technology owners who have filed for patent are from Asia. Also, Republic of Korea and China are leading patent fillers for climate technologies in the recent years.

Interest in sustainable technology innovation is increasing. As per a research conducted by McKinsey, if the potential of energy in land, water, and material is tapped with the use of existing technologies, it would meet almost one-third of the total demand in 2030. The emerging economies are expected to contribute around 80% to these prospects. If the projection is achieved, the saving would amount to $2.9 trillion in 2030.

7.3. TECHNOLOGICAL DEVELOPMENTS FOR GREEN MANUFACTURING

Technology plays a crucial role in solving environmental problems around the globe. There has to be drastic technological changes to mitigate the environmental impact due increases in production and consumption. Innovation in technology is aimed at changing the recycling methods, producing alternative non-toxic materials, changes in production process, efficient use of scarce resources and control emission of toxic gases (Figure 7.2).

Figure 7.2: Green technology.

Source: https://commons.wikimedia.org/wiki/File:Researching_green_technology.JPG

7.3.1. Sustainable Energy Technologies

The energy technologies can be studied under two broadheads

- Green technologies for energy-intensive business.
- Green technologies for less energy intensive manufacturing business.

Technologies adopted by some energy-intensive industries are discussed below:

7.3.1.1. Plants Manufacturing Iron and Steel

Steel manufacturing comprises of various processes. The steps of the processes are not rigid, and the combinations depend on energy supply, resource availability, product mix. The three chief processes routes are discussed below:

- **Blast furnace route:** Coke and iron ore are the chief raw materials in this route. Pig iron produced by combining the two are then in a blast furnace is then converted into steel in a basic oxygen furnace. Extensive heat and pressure are used in the process along with coke which makes this an energy consuming process.
- **Scrap route:** This process does not involve coke and iron making process and is mainly with scrap for iron input. Hence this route is less energy consumption compared to BOF.
- **Direct reduced iron route:** In this route, iron ore is the main component. The size of the manufacturing unit, the fuel used, and the properties of the iron ore determine the amount of energy used in this process.

In the recent times, there is more emphasis on reducing the heating and melting process in the blast furnace route.

The following Table 7.1 lists Greenhouse Gases reduction and Energy saving measures in the Iron and Steel industry:

Table 7.1: Measures for the Iron and Steel Industry to Reduce Greenhouse Gases

Coke Dry Quenching	Pulverized Coal Injection	Direct Rolling (Integrated Casting and Rolling)
Natural Gas Injection	BOF Heat and Gas Recovery	Improved Ladle Preheating
Non-Recovery Coke Ovens	Hot DRI/HBI Charging to EAF.	Variable Speed Drives on Flue Gas Control, Pumps, and Fans
Coal Moisture Control	Variable Frequency Drives on Ventilation Fans	Coke Stabilization Quenching
Top Pressure Recovery Turbines	Blast Furnace Process Control	Foamy Slag Practices

Plastic Waste Injection	Emissions Optimized Sintering	Waste Heat Recovery in Sinter Plant
Thin Slab Casting - Near Net Shape Casting	Increased -Blast Furnace Top Pressure (> 0.5 Bar Gauge)	Endless Strip Production (ESP)
Strip Casting – Castrip® Process	Stabilization Quenching	BOF Bottom Stirring
Post Combustion of EAF Flue Gas	Additional Use of Coke Oven Gas	Scrap Preheating Coke
Heat Recuperation from Hot Blast Stoves	Hisarna	Software Tools to Boost Steam System Efficiency
Improved Recovery of Blast Furnace Gas	Direct Current (DC) Arc Furnace	Optimal Charge Calculation in EAF

Source: https://www.unido.org/sites/default/files/2015-01/Institute_Emerging_green_trends_Future_of_Manufacturing_0.pdf

7.3.1.2. Plants Manufacturing Cement

Cement production mainly uses four major process routes, namely dry, semi-dry, semi-wet, and wet. The raw material available determines which technology would be more appropriate. Dry process is less energy-intensive compared to other routes.

Developed countries mostly opt for dry processes since dry materials can be accessed easily. Manufacturers are moving towards adoption of dry processes for setting up new units or up-gradation and expansion of existing ones.

However, some countries like Australia and New Zealand are still using wet processes which are energy-intensive. Those countries are making efforts to gradually shift to dry process.

The main of cement is clinker which is obtained by heating limestone and clay at a high temperature. This process uses a lot of energy and also emits high amounts of CO2. All the processes involved such as crushing and grinding the inputs are energy-intensive.

Recently technologies have been developed that cut-down the energy consumption to a great extent and can also limit the amount of CO2 that is emitted. It can be categorized into the following heads:

- Adoption of technologies that are energy efficient.
- Use of raw materials and fuels that can reduce the environmental impact.
- Use other blends to limit the use of clinker.

Substitutes for cementations materials, carbon capture, and storage are some other alternatives that are being explored.

CELITEMENT is a new process that has been developed with the objective of reducing energetic emission and process emission. CELITEMENT and other similar processes involve much less heat and use alternative chemicals which have less environmental impact. These new processes aim to cut down the emissions by almost half of what is emitted in the current process.

There are some challenges with respect to the above methods. First, the processes are yet to be tested for higher volumes. Second quality assurance certificates are yet to be obtained before using them for constructions.

These processes need pervasive changes in the processes, and it will take at least another five years for these changes to take place. The pioneers in this field will enjoy huge benefits in terms of savings in raw material cost and increased efficiency in production.

The following Table 7.2 lists Greenhouse Gases reduction and Energy saving measures in the Cement industry:

Table 7.2: Measures for the Cement Industry to Reduce Greenhouse Gases

Cement grinding with Horomill	Limestone Portland Cement	Use of EAF slag - Cem-Star®
Carbide Slag as Raw Material	Kiln Shell Heat Loss Reduction	Kilns Cement with Poz-zolana
Preventive Mainte-nance	Addition of Pre-Cal-cination to Kilns with Preheaters	Vertical Roller Mills for Finish Grinding
Replacing Vertical Shaft	Conversion to Recipro-cating Grate Coolers	Combustion System
Efficient kiln drives	Improved Burnability Using Mineralizers	Variable Speed Drives

Optimizing Fuel Properties	Waste Oil and Oil Sludge as Fuel	Refuse Derived Fuel (RDF) Coprocessing
High-Efficiency Fans for Preheaters	Gravity Type Blending Silos	High-Pressure Roller Press
Emerging Grinding Technologies	Dry Kilns with Multistage Preheaters and Pre-calcination	Improvements Bucket Elevators for Kiln Feed
Cement grinding with Horomill	Geopolymer Cements	Waste Heat Recovery for Power Generation
Conversion of Long Dry Kilns to Preheater/Precalciner Kilns	Revolving disc clinker cooler	High-Efficiency Roller Mills
Variable Speed Drive and High-Efficiency Fans for Dust Collection	Proper Sealing and Seal Replacement	High-efficiency motors & drives
Low-Pressure Drop Cyclones for Suspension Preheaters	Used Tyre Gasification Emerging	High-pressure roller presses as pre-grinding to ball mill
Grinding Technologies		High-efficiency Separator/ Classifier for Coal Grinding
Variable Frequency Drive for Clinker Cooler Fans		Process Control and Optimization in Clinker Making

Source: https://www.unido.org/sites/default/files/2015-01/Institute_Emerging_green_trends_Future_of_Manufacturing_0.pdf

7.3.1.3. Plants Manufacturing Paper

The processes that are energy intensive in paper production are the steps involved in producing pulp and dry paper. The manufacturing plants are either standalone pulp mills or integrated mills that produce both pulp and paper. Pulp mills consume more energy compared to integrated mills. The most popular chemical pulping process is the Kraft pulping since the quality of output is high and is used in high-grade quality papers. This process uses harsh chemicals and high amount of heat. Kraft mills are self-sufficient in terms of energy. They generate most of the energy while producing by-products. Another process called sulfite pulping exhibits a similar characteristic, wherein most of the energy requirement of the plant is met from energy generated during production of by-product.

Mechanical pulping is an energy efficient process. It produces high yield and requires less energy overall. Thermo-mechanical pulping is an application of the process where heat is recovered at different grades. This process results in reducing the overall energy consumption. However, this technology requires electricity as its chief source of energy and hence primary requirement of energy is generally high, and so is the amount of CO_2 emitted.

One good prospect for conserving energy in paper production is by using the recovered papers to produce pulp. This process is more energy efficient and also emits less CO_2 as compared to using virgin material. The estimated savings of energy by adopting this method is as high as 35 percent. It is important to note that recovering pulp is not an easy process and entire paper production lifecycle has to undergo a transformation.

The quality of pulp and grade of paper used in the production determine the energy requirements of the machineries. The amount of energy consumed varies from one mill to another based on the above factors. This is the reason why integrated mills are much more efficient in consumption on energy as they adopt better processes and the intermediate pulp is reused for production.

Combine heat and power (CHP) is an alternative emerging technology that reduces the energy consumption of this industry. The industry's CHP potential is estimated to range between 0.3–0.6 EJ/year. It is estimated that CHP can result in huge savings in terms of energy and fuel. If the conventional technologies are replaced by CHP, it will amount to almost 20 percent savings in fuel and 3 percent savings in energy.

According to the International Energy Agency, the following concepts and techniques can improve energy efficiency in the manufacturing plants and limit CO_2 emissions.

- **Black liquor gasification:** This concept aims to convert the black liquor mass into a source of energy and makes the pulp making process more resourceful. This process also recovers some pulp chemicals from the liquor mass.

- **Bio-refinery:** This concept suggests that biomass should be used instead of other fuels to produce energy. The biomass conversion requires the use of special machineries to produce heat, electricity, and useful chemicals from the waste products.
- **Recycling paper:** This process deals with recycling paper and makes it fit for reuse. Scraps produced in the paper mill, discarded material, and other waste papers can be recycled using a separate set-up.

A paper mill conventionally uses various types of woods and different blends of inputs to manufacture several types of paper or pulp. It is easier to obtain the energy requirements of the various products but the cumulative consumption at the end of the year depends on the amount of paper or pulp produced under each category. The sub-processes involved in each process and the different techniques of production make it difficult to set a benchmark for the paper plants. However, benchmarking can be done for certain paper mills that are uses particular type of paper and pulp and have the similar production processes and can be compared.

The methods adopted to make the production process efficient may also have impact on the product, so it is important to understand how it alters the paper strength, opacity, and other characteristics.

The following Table 7.3 lists Greenhouse Gases reduction and Energy saving measures in the Paper industry:

Table 7.3: Measures for the Paper Industry to Reduce Greenhouse Gases

Combined Heat and Power (CHP) Generation	Continuous Digester	Digester Blow/Flash Heat Recovery
Improvements in Chemi-Thermomechanical Pulping (CTMP)	Black Liquor Gasification	Waste Heat Recovery from Paper Drying
Using Drum Pulpers	Heat Recovery in Thermomechanical Pulping	
Optimization of Water Removal in Forming	Optimization of Water Removal in Press Section	Condebelt Drying
Dry Debarking	Solids Concentration	High-Temperature Odor Free Recovery Boiler

Increased Use of Re-cycled Pulp	Low Consistency Re-fining (LCR)	RTS Pulping Black Liquor
Advanced Thermo Mechanical Pulping (ATMP)	Batch Digester Modifi-cations	Replacing Pneumatic Con-veyors with Belt Conveyors

Source: https://www.unido.org/sites/default/files/2015-01/Institute_Emerging_green_trends_Future_of_Manufacturing_0.pdf

Technologies adopted in some less energy-intensive industries are discussed below.

Electric motors are used to supply power to the machines since they can generate electrical power from mechanical power. A manufacturing unit typically has numerous handling and processing functions like pumping, conveyance, compression for which it utilizes electric motor-driven systems. The electrical motors and their control consume a significant portion of the electricity in a motor-driven system.

However, they do not affect the overall energy efficiency of the system. The other accessories of the system have an impact on the amount of mechanical energy required as well as the loses that can be attributed to the delivery of the power. System parts like pipes, fans, and ducts fans collectively impact the overall energy consumption.

Therefore, it is vital to view the components of the motor-driven systems as independent and interacting parts so as to improve the efficiency of the system with respect to energy consumption.

The competence and effectiveness of a motor-driven system depends on two factors: first, the kind of advanced solutions used and the degree to which they are used, second, the design of the various elements of the system. A motor-driven system can be made more fruitful towards energy consumption by ensuring that:

- The motors that are used in the system consume optimum amount of power.
- The principal parts selected are appropriate and productive.
- Select the best design for the entire system that can make the operations energy-efficient.

It is relatively easy to evaluate how efficient an electric-motor is and to identify areas for improvement. For assessing efficiency, class codes have been introduced namely IE1, IE2, IE3, and IE4. It is a universally accepted classification where IE is an abbreviation of International Energy Efficiency Class. Energy savings around 5 percent is possible by using the highest quality of motor.

The various system components and processes adopted have a significant impact on energy conservation. This poses a challenge in setting benchmark for the entire system.

The following Table 7.4 lists Greenhouse Gases reduction and Energy saving measures in the Motor Industry:

Table 7.4: Measures for Motor Industry to Reduce Greenhouse Gases

High-Efficiency Motors	Predictive Maintenance Program	Variable Speed Drives for Pump Systems
High-Efficiency Motors in Pump Systems	Variable Frequency Drives	Installing Zero-Loss Condenser Drains
Pressure Profile Optimization - System Controls		

Source:https://www.unido.org/sites/default/files/2015-01/Institute_Emerging_green_trends_Future_of_Manufacturing_0.pdf

7.3.2. Sustainable Climate Technologies

The climate change technologies in the manufacturing sector can be examined under the following broad heads:

- Carbon capture and storage technologies used in manufacturing units.
- Use of smart grids in renewable technology.
- Technologies specifically designed to control the emission of harmful gases.

7.3.2.1. Carbon Capture and Storage in Manufacturing Units

Carbon capture and storage/sequestration (CSS) refers to the technology that can trap the CO2 emitted at the source and prevent its entry into the atmosphere. This technology is used at the source or point of emission of the gas and is transported and deposited in a secluded location which is usually an underground storage from where it cannot pollute the atmosphere. These emissions could be generated from fossil fuels or electricity generation. It is a product of geo-engineering.

It is mandatory in steel production and other manufacturing where there is continuous usage of fossil-fuel based reducing agent.

This is an emerging technology, and a lot of research is still going on how to make effective use of this technology. Industries which are not energy-extensive like cement, steel, paper can adopt CSS. CSS requires significant amount of energy. It is estimated that the overall potential of emissions reduction in the iron/steel industry by using CCS ranges between 0.5 Gt to 1.5 Gt of CO2/yr. CCS for blast furnaces cost around US$40/t CO2-US$50/t CO2. CCS for direct reduced iron can cost less than US$25/t CO2. It is important to note that this technology is expected to make a significant impact only by 2030, on making the manufacturing process more environments friendly.

The European Union (EU) and European Steel Platform have rolled out the ULCOS project to carry out advanced research to bring improvements in manufacturing process and reduce emission by half.

CSS is the most preferred method for reducing the CO2 pollutants in the atmosphere. CO2 can be prevented by using the following technologies:

- Pre-combustion capture;
- Oxyfuel combustion;
- Post-combustion capture.

Oxyfuel combustion and post-combustion capture are very pertinent for the cement manufacturing units. There is an ongoing research on various areas where the use of carbon capture technique will be pragmatic. Technologies exists which facilitate transportation and sequestration of CO2; however, the scope for application are few. This technique is yet to be tested on a larger scale of manufacturing. Also, the cost for implementing this technology is high, so the policymakers have to intervene and make it mandatory so that CCS finds acceptance in the manufacturing sector.

The International Agency has prepared a comprehensive technology guide for application of CCS in manufacturing. With regard to industrial applications of CCS, the roadmap states: "Whereas the power sector can take advantage of alternatives to fossil fuels, in several industries deep emission reductions can only be achieved through CCS [...] For developing countries, CCS could be part of a low-carbon industrial development strategy. If CCS can be implemented through the United Nations Framework Convention on Climate Change (UNFCCC), Clean Development Mechanism (CDM) or other new global climate mechanisms, the cost barrier could be partly overcome. It is likely that if CCS moves forward under the CDM, the first projects will be in industry."

7.3.2.2. Smart Grids and Renewable Energy

The Imperial College London published its finding regarding sustained use of smart grids for renewable energy. As per the report "that these Smart Grid advances may enable sub-Saharan African countries to leapfrog elements of traditional power systems in terms of both technology and regulation. This could accelerate national and regional electrification timeframes, improving service delivery, minimizing costs and reducing environmental impact."

The report also introduces "the notion of Just Grids to reflect the need for power systems to contribute towards equitable and inclusive global, economic, and social development. While Smart Grids may provide an efficient mechanism to address the massive electricity infrastructure building requirements, Just Grids will help guarantee access to modern energy services without marginalizing the poor."

Smart grids in renewable energy can be applied in the manufacturing sector but not as a long-term fix. The entire process of manufacturing has to be restructured before adopting renewable energy technologies. This is not a conventional source of energy; hence manufacturing process needs modifications and novelties so that this technology can be applied to the industry. Some of the alternative sources of energy are:

- **Wind Energy** can be used as an alternative source of energy. We can the example of Lafarge Company in Morocco. The electricity rates are high in Morocco and to manage the high costs and maintain stable supply, the company uses wind energy.
- **Solar Power Systems or Photovoltaic System** have a wide scope globally, for application in manufacturing industries. Manufacturing processes with medium temperature range requirements can use small concentrated solar power systems. For low-temperature heating and cooling requirements in office spaces and manufacturing processes, performant solar collectors can be used. Emerging designs of these systems can generate thermal energy up to 400.c.

The potential for using renewable energy is more in case of small scale sectors that are energy efficient like food industry, textile industry. The challenge with respect to using this technology is that it is capital intensive and hence does not have widespread acceptance. It is crucial to break this vicious cycle (Figure 7.3).

Figure 7.3: Solar power grid.

Source: https://www.flickr.com/photos/viucsr/6271623794

- **Biomass** as a renewable source of energy is versatile and can replace the conventional sources of energy. It can meet the heating requirements of processes needing low as well as high temperature. It is a good substitute for fossil fuel and can meet localized energy requirements. Biomass is the most preferred alternative source of energy due to various reasons. First, the cost of generating energy is low. Second, the energy produced per unit of mass is high. Third, there are various choices with respect to transportation of biomass including barge and pipeline, trucks, trains, etc.

Other Climate Technologies:

- **Waste-Water Electricity Generator:** The Oregon State University has invented a hybrid electricity-generator. This technology microbial fuel cells and reverse electrodialysis to generate electricity using waste water. The generator produces energy for self-sustenance as well as excess that is contributed to the main power grid. If this technology is developed on a large scale, it will form the foundation for energy-water sustainability.
- **Molten Salt Storage:** It can be used in storing solar energy for future use. The salts can absorb high amount of heat and store them during the day which can be used to generate steam, run

electric turbine even in the absence of sunlight. This makes solar energy a feasible alternative to conventional sources of energy.

- **Artificial Photosynthesis:** Research is going on to develop a technology that can produce energy by combining sunlight and carbon-dioxide. This technology will eliminate CO2 from the atmosphere on one hand and produce renewable energy on the other hand. On a small scale, this technology has been proved, but the challenge lies in scaling it. The barriers to this technology are captured carbon storage and converting solar energy. Overcoming them would ensure the success of this technology.

7.3.2.3. Technologies Specifically Designed to Control the Emission of Harmful Gases

Technological advancements in have facilitated manufacturing concerns to rationalize the production process and maximize earnings. But only in the recent times, technology is being designed to make manufacturing eco-friendly.

Globally, regulations are being enforced to control the emission of harmful gases from manufacturing units. Many of this emission include Hazardous Air Pollutants (HAPs) and Volatile Organic Compounds (VOCs). Smog and low-lying ozone can be attributed to emission degrade bin the presence of sunlight. The best method to treat these hydrocarbon-based pollutants is by using thermal and catalytic oxidizers.

Various abatement technologies are being applied on various manufacturing processes to control process emissions. Depending on the process conditions, destruction requirements and energy demands at a given facility; flares, vapor combustors, thermal oxidizers, and catalytic systems have been applied for emission destruction.

Environmental personnel in the manufacturing units apply environmental catalyst for controlling emissions of VOCs and HAPs in the gases emitted during production processes. These catalysts promote oxidation of these gases, that too at lower temperatures making it an energy efficient process.

The only downside to this technique is that catalytic oxidizers are expensive. Factors like maintenance concerns, higher capital cost, potential for higher operating expenses and subsequent greenhouse gas emissions is forcing the industry to find an alternative. Another abatement technology used widely is the Regenerative Thermal Oxidizer. It is a very efficient pollution control application. Almost 97% of the thermal energy from burning of fossil fuels is used by RTO to treat the pollutants emitted.

The RTO technologies have evolved, and now it can be established for process exhaust gases with high concentration. Hot Gas Bypass is a component in RTO which can handle challenging conditions like spikes in the emission.

The RTOs have a low maintenance and replacement cost since they do not incorporate catalyst into the design. Also, the absence of catalyst negates the risk of degradation in performance efficiency over the RTOs operational life. Only selected parts of the RTO have to be replaced like fan belts, thermocouples, and flame sensors. The valves are very sturdy and can endure over million cycles. Usually, they function for years before they start degrading.

The RTOs have a clear edge over catalyst oxidizers since the usage of supplemental fuel is lower when exhaust concentration in the emissions are low. It does not need auxiliary fuel when concentration is above 3% LEL (lower explosive limit) as it is mostly self-sufficient. This has a direct correlation on greenhouse gas emissions from the auxiliary fuel combustion process; mainly CO2 and NOX are also reduced compared to older emission control designs with the higher fuel usage. For instance, large scale chemical plants can save hundreds of dollars per hour on auxiliary fuel by establishing RTOs since they are highly thermal efficient.

Abatement systems are an integral part of manufacturing process since non-compliance may often result in shut-down of operations or even closure of the facility. Also, it is important to note that the technology has to be used effectively to avoid drastic impact on operating expenses and earnings.

It is important to remember that technology alone cannot transform production patterns, but a shift is also required in models, ideas, organizations, and societal establishments. The role of technology cannot be undermined, and the efficient use of technologies available in the market and innovation of new technologies is essential for ensuring sustainable production and consumption.

7.3.3. Laser Technology

The astounding rise in demand for goods, manufacturing companies, has the burden of meeting the demand for these goods while adopting eco-friendly processes at the same time. Scientists working in the areas of laser and photonics recognize this urgency to adopt environmentally benign manufacturing practices that are also cost-efficient and maintain quality.

Emerging economies in particular face unique infrastructure challenges since sharing resources is vital. Government representatives, academic person, have done researches in the field of laser technology that can help the emerging economies move forward towards meeting sustainability goals. Projects such as manufacturing stronger steel can ensure longevity of structures, simplifying the production of graphene and producing titanium that can be easily shaped will make the process energy efficient. When the durability of a material is increased, achieving sustainable goals become easier in the developing countries.

Stainless steel widely used material in constructions is very sturdy and corrosion resistant in dry conditions but cannot survive long in gaseous and wet environments. Laser technology can play a huge role in making this material more durable. The laser process is an eco-friendly process and has many advantages (Figure 7.4).

- It reduces the waste material generated;
- The process is energy efficient;
- It eliminates hazards of radiation;
- The technology can process complex shapes and high-quality products.

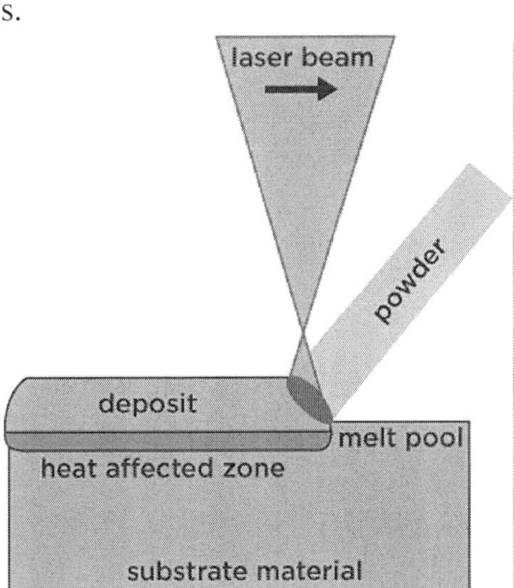

Figure 7.4: A schematic of laser surface alloying to make stronger steel.

Source:https://spie.org/membership/spie-professional-magazine/archives/2017_july_archive/sustainable-lasers-in-africa?SSO=1

The laser technology is still in its infant stage, and the manufacturing sector has to collaborate with the scientists under how a laser processed material performs.

Titanium is the most preferred material in marine, aerospace, biomedical, and chemical sectors because it is sturdy, has less chemical reactivity and is biocompatible. Scientists are exploring ways to reduce it cost and simplify its manufacturing process.

Laser beam can shape metals accurately by heating a localized area fast. It is more energy-efficient than the traditional methods of heating since the operation is localized. It also takes less time and offers better control over the power source.

The laser technology can be controlled better than mechanical force methods. Extensive research is being carried out, and the results are positive and suggestive that the laser technologies can transform the manufacturing processes drastically in the future. The laser technology is expected to make materials cheaper, energy-efficient, reusable, and environmentally benign.

The drawback of laser technology is that it is capital intensive and requires skilled labor for installation. The research is also not adequately funded making innovations and progress difficult. Leveraging this research to develop SM, and pushing techniques to widespread use, requires a commitment to sharing resources and knowledge across and within international boundaries. It is challenging to even procure the elementary equipment at research institutes. Many of these equipment belongs to the colonial era and do not function properly.

It is important that policymaker realize that applying laser technology will be a big step towards achieving SM and invest in R&D for development in this field.

7.3.4. Clean Technology vs. End-of-Pipe Remedies

In the past, emphasis was more on 'cleaning technologies' than clean technologies. Technologies were used as add-ons to reduce the environmental impact. They were called end-of-pipe technologies or control devices. Clean technologies are those which modify the production processes, raw materials that go into production and end product so that they are environment-friendly. Clean technologies are superior to end-of-pipe technologies because they prevent toxic materials from being generated in the first place. The products are more environmentally benign.

The technologies used in the processes should be efficient in terms of usage of resources and energy and reduce emissions and toxic wastes. For instance, alternate drying techniques can help save water, process-integrated flue-gas cleaning and filter systems can reduce generation of wastages. Processes should be transformed so that eco-friendly materials like solvent free inks, metal-free pigments can be used. The end product should also be environmentally benign and should have the least environmental impact both during manufacture and usage. Also, the waste generated should be reused in the process than being disposed.

According to the Organization for Economic Cooperation and Development (OECD), while manufacturers were investing in end-of-pipe technologies to limit the environmental impact, the investment in clean technology was only 20 percent. There are primarily three reasons for this. First, cleaner technologies can be harder to procure. Second, manufacturers prefer to use old technologies till they are useful instead of replacing them while they are in operating condition. Third, manufacturers tend to avoid structural changes and are more comfortable with the existing processes than having to adapt to a new one.

End-of-pipe is an attempt to fix a problem rather than prevent it from happening. Hence it can lead to other challenges. For instance, available carbon is converted into CO2 to improve the thermal efficiency of fossil fuels. As a result, CO2 is a Greenhouse gas is emitted as collateral damage. Also, this process increases the temperature of the combustion releasing more nitrogen oxides in the atmosphere. Thus, other less desirable reactions occur in trying to make combustion an energy-efficient process.

Barry Commoner, an American cellular biologist, says, "if technology is indeed to blame for the environmental crisis, it might be wise to discover wherein its 'inventive genius' has failed us and to correct that flaw before entrusting our future survival to technology's faith in itself."

All technological alternatives are seldom explored. One of the chief causes is that alternative technology tends to be costlier. There are other reasons too and the hesitation to adopt new technologies can be partially discussed in terms of technological paradigms. It takes more than the existence of appropriate or clean technologies to ensure their widespread adoption.

7.4. PRIORITY AREAS FOR ACTION

International Renewable Energy Agency conducted techno-economic analysis for the usage of alternate sources of energy and came up with four areas that warrant attention from policymakers and industrial stakeholders. These are:

7.4.1. Energy-Intensive Sectors

These sectors consume almost three-fourths of the total requirement of energy of the industry. Also, these manufacturing units are expected to operate for a long time. So, it is crucial that the energy-intensive sectors make alternative sources of energy, an integral part of energy source not just for their upcoming facilities, but also the existing ones.

7.4.2. Small-Sized and Medium-Sized Enterprises (SMEs)

Almost 90% of the businesses in manufacturing sector are SMEs. Hence, they play a vital role in making renewable technology of energy a success by increasing its deployment rate. In the long run, they help in cost reductions and increase local opportunities in the manufacturing sector.

7.4.3. Electrification

Of the total energy use in manufacturing sector, almost 20% is electricity. There are manufacturing industries that are electricity-intensive, like plants manufacturing non-ferrous metals. So, the electricity consumption of the production process is high. It is important to decarbonize power sector by increasing other alternatives for renewable energy of the electricity sector.

It would be a good strategy if the electricity-intensive industries are relocated within a close range of these renewable energy power plants. Self-sustaining and energy-efficient plants that are powered from sources like wind turbines and solar power units are being used in commercial manufacturing to conserve electricity.

7.4.4. Regional Aspects

Regional potential depends on production growth, ratio of existing and new capacity, and renewable resource availability. Policies that support the manufacturing facilities using renewable energy have to be formulated. Pricing has to control for energy and policies to address the threat of climate change has to be formulated to ensure a level playing field.

Globally, manufacturing units around the world are deploying alternative sources for generating power using smart grinds so that energy is conserved, and productivity is increased. Telsa Motors has its manufacturing unit in Sorey County, Nevada, and is breaking ground by switching to alternate sources of energy. It will now entirely operate on power generated from PV systems, a wind-mill, and a closely located geothermal electricity plant. Method Soap is running a platinum factory that is LEED- certified. It has 23-storied wind turbine and solar photovoltaic collectors on-site from which the energy requirements are met. These alternative sources of energy are a major contributor towards reducing the ecological impact in manufacturing.

By 2030, it is estimated that renewable sources of energy would contribute to almost 30% of the total final energy consumption for global manufacturing. The pre-requisites are:

- Availability of low-cost and sustainable biomass sources;
- Reduced capital costs for emerging technologies;
- Implementation of potentially around 34% of carbon-emission pricings.

Across the globe, the manufacturing sector is increasingly achieving innovation, expansion, and strategic advantage in clean energy manufacturing by developing and implementing these alternative sources.

7.5. IMPACTS OF EMERGING GREEN TECHNOLOGIES ON INNOVATION IN MANUFACTURING SECTOR

The green technologies have evolved over the last two decades. There are pronounced differences in the role innovation plays within the different strategies (Figure 7.5).

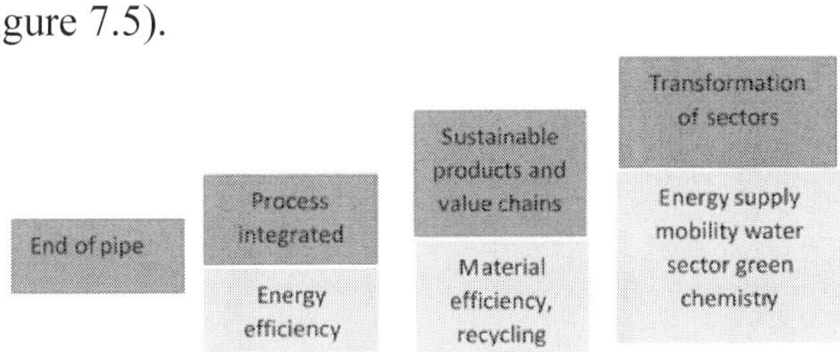

Figure 7.5: Innovation dimension of environmental technologies.

Source: https://www.unido.org/sites/default/files/2015-01/Institute_Emerging_green_trends_Future_of_Manufacturing_0.pdf

In the initial phase of implementation of green technologies, end-of-pipe solutions were employed. These were add-on techniques that did not alter the processes in any way. It had no impact on the production's core business. Innovation pertained mainly to curtailing emissions and reducing costs. This approach was responsible for curbing CO2 pollutants.

In the next phase, strategies were more integrated with the processes. These technologies aimed to reduce energy consumption thereby limiting emissions. This approach demanded more integration of the core business process into these techniques. Thus, aspects such as the quality of production and changes in economic productivity become relevant. The innovations had to incorporate these aspects also, making it a challenging task. At the same time, it was rewarding because these innovations could also make the production process more cost-efficient and make the production capital more advanced while simultaneously reducing environmental impact. We can take the example of steel factories or electric motor manufacturing which has undergone a transformation in the manufacturing processes to pave way for green technologies.

In the recent times, aspects such as closing of material cycles and integration of product policy and product use have gained importance. The areas of concern are the use of non-toxic material for products, utilization of scarce resources. New tools for assessments have come up, like the lifecycle assessment. These strategies address the environmental issues along the entire value chain and hence are more complex. It is not just concerned about the processes within the factory, but also the processing of input material and recycling of the product after end-use.

Introduction of new products also require new models of production which necessitates change in the processes. These strategies give rise to new business prospects as new markets can be penetrated only through innovation in products. In conclusion, it can be said that the distinction between environmental strategies and innovation is getting blurred, and they now appear to be the same side of a coin.

There is universal agreement over the fact that innovations are the primary drivers for reducing ecological impact and propelling the economy forward. Advancement in the field of innovation, therefore, garners attention. To solve any technical or organizational problem, innovation is the stepping-stone. Innovations is not only about finding new solutions but also improving current solution. Although the objective of the innovation may not always be to reduce environmental impact but should contribute in some way to the cause of environment. The term sustainability technologies are used in this context as well to highlight the economic dimension of eco-innovations with regard to integration into the production process, new processes or even the transformation of sectors. Invention, technological development, and application are not always a linear process and have incorporate feedbacks to bring improvement. Producers of the innovation will look for social acceptance of the product and will try to receive feedback from the market. Hence it is important to understand out how and why new ideas proliferate.

7.6. ADVANCES IN GREEN MANUFACTURING PRACTICES

7.6.1. Country Specific

Various green manufacturing practices and initiatives have been taken by different countries, some of which are discussed below:

- **European Union (EU):** The EU has issued directives with respect to green manufacturing to all its member states. The directive is for guidance, and the states have formulated their own policies and implement them. EU has laid more emphasis on design for environment (DFE), elimination of toxic wastes, end-of-life (EOL), waste electrical and electronic equipment (WEEE).
- **Germany:** The 'Green Movement' was initiated by Germany. Some of the steps Germany has taken include eco-labeling, green taxes. Germany advocates the use of the most advanced technologies available and is very stringent with respect to green manufacturing. It follows the principle of anticipatory protection, i.e., works towards identifying future environmental impacts and tries to prevent them. The country believes that investment in environmental causes will have a positive effect on productivity.
- **USA:** The manufacturing units are adopting the GM philosophies rapidly. The manufacturers are paying attention to all aspects including efficient use of energy, reuse of packaging, implementation of ISO 14001 and recycling of water. Multinational companies are also accepting EU directives like WEEE and ELV alongside practicing conventional methods.
- **Japan:** Japan has also taken various initiatives like development of lead-free solders, emphasis on recycling, along with adoption

of ISO 14000. It has also formulated guidelines for purchase by government agency, has placed limitations on usage of hazardous substances, hybrid cars. Japanese have it in their culture to conserve scarce resources and avoid wastage; hence this country has taken a giant leap towards green manufacturing.

- **China:** has implemented laws relating to information electronic production pollution control and management. WEEE directives will also be rolled out shortly. Among all industries, iron industries, in particular, have adopted green manufacturing practices and are reaping their benefits.
- **India:** The manufacturing companies have taken various initiatives in the recent times like efficient use of energy, judicious use of water and other resources, limit the use of hazardous materials, management of wastes and minimize emission of dangerous gases. Green purchases have also found acceptance among manufacturers. The leading manufacturers are making efforts to getting ISO 14001 certified. They are also trying to adhere to the Restriction of Hazardous Substances (RoHS) Directive.

In conclusion, it can be said that Germany focuses more on eco-friendly recycling of wastes and regulating the characteristics of final product. USA on the other hand predominantly emphasizes on regulating the production. Japan considers the long term impact of its strategies and incorporates GM into their business plans including the launch of new products to increase the market, conserving resource. Europe is heavily concerned with product EOL, supply chain and reverse logistics (RLs).

7.6.2. Company Specific

Multinational companies have also started realizing their social responsibility and have moved towards Green Manufacturing. Initiatives taken by some of the brands are discussed below:

- **Toyota:** A leading car manufacturer has developed a purchasing guideline for its suppliers as per the ISO 14000 standard. It has also adopted a policy to measure the energy consumed by the machineries at various stages of production. The company also emphasizes on recycling waste with minimum impact on the environment.
- **Mercedes Benz:** One of the chief corporate objectives of Mercedes-Benz is protecting the environment. It has implemented

the recycling system MeRSy. This technology recycles 97% of the production waste. The Mercedes-Benz Environmental protection UK stated that it uses only 21kg of landfill for managing waste of each vehicle.

- **Apple:** Numerous initiatives have been taken by this iPhone manufacturer. Innovations have helped to optimize the energy consumption of their hardware tools. They have launched take-back programs in Europe, U.S. Japan, and Canada and have diverted almost 34 million pounds of electronic equipment from landfills globally.
- **Samsung:** This electronics manufacturer has laid down detailed plans to curb environmental impact. It aims to limit the Greenhouse gas emitted by its production facilities by half in the next five years. Also, it plans to curtail toxic wastes by 84 million tons in the same period. It has allocated US$ 4.35 billion for eco-management initiatives and also augments Green partnerships with vendors and other partners. It has mandated ISO 14000 compliance for all its supplier facilities and has a dedicated audit team that checks for the compliance in the supplier's facilities.
- **Ford:** has designed a car expressly for European take-back. It designed the Ford Escape Hybrid, which has been rated by Environmental Protection Agency as the most fuel-efficient car globally. This pushed its sale for hybrids.
- **General Motors:** has made it mandatory for all its Tier-one suppliers to be ISO14001 certified since 2001. The vendors have also received a list of restricted materials that cannot be used in the manufacturing of components. The vendors also worked on a World Energy Council project to improve their strategic position and sustainability performance at the same time ensure savings. One such measure was to reduce the weight of the truck manufactured which made the vehicle more fuel-efficient.
- **Motorola:** This telecommunications company has taken various steps to ensure recyclability, minimize the use of packing materials, use of plastic in labeling, limit use of toxic inputs and conserve energy. It has also undertaken supplier training programs and supports Green purchasing. It has established a lab in China to restrict the usage of toxic materials and screen consignments from vendors.

- **Philips:** is a leading manufacturer in healthcare, consumer lifestyle, and lighting. In the recent times, the company has shifted to processes that are energy-efficient. It is also limiting the use of scarce resources, has adopted technologies that takes care of lead solvents and decomposition products, has adopted mass balance practices.

Globally manufacturers are striving to embark on the journey of sustainable production. They are bringing innovations in the products so that they are less toxic, making manufacturing techniques environment-friendly, and production processes energy-efficient to ensure sustainability in the long run.

7.7. CONCLUSION

Achieving manufacturing sustainability is a challenging task and also a priority. The entire manufacturing sector has to undergo a change. Significant structural changes are required in the manufacturing industry. Resource base has to be shifted particularly in the field of water, energy, chemistry, and transportation. Extensive innovations are required not just in the manufacturing process, but also allied processes. Co-evolution of technology is required in the field of education and skills development. Thus, complexity of innovation goes beyond the micro-level in economy and affects the macro-level to an extent.

Environmental strategies are multifaceted with respect to innovation. The consequences for green technologies in manufacturing are significant. Green technologies are suited for both energy-intensive and energy-extensive manufacturing sectors. Cross-cutting industrial technologies are efficient for conserving energy. CCS technologies bear similarities to end-of-pipe technologies. Smart grids in renewable sources do not alter the manufacturing processes much, but they are energy efficient and lead to cost reductions while also maintaining the required supply.

Whether technology is a boon or bane depends on how it is used or abused. Technology has to be applied with due precaution so that they remain environmentally benign. Technologies may sometimes have unintended consequences. For example, plastic was manufactured to be persistent, but it does not degrade and makes the environment toxic. Technology can coexist in the ecosystem "if its aims are directed toward the system as a whole rather than at some apparently accessible part."

Technologies are integral for SD. But it is important that the policymakers take steps to force fundamental technological changes than just end-of-pipe solutions that prevailed for long. The industry needs to take a long term perspective of these changes and be ready to bear the financial burden in the short-term and evolve accordingly.

One has to remember that implementing clean technologies will not be sufficient to bring down pollutions to an acceptable level since manufacturing is expected to grow exponentially. It is required for toxic discharges and waste flows to reduce drastically to achieve the objective of making life on Earth Sustainable.

The increased reliance on technology alone will not enable humans to maintain their lifestyles and ensure an increased level of consumption for everyone. The need of the hour is a radical social change and change of priorities.

REFERENCES

1. Alcorta, L., (2018). *Emerging Green Technologies for the Manufacturing Sector*. [eBook] Available at: https://www.unido.org/sites/default/files/2015–01/Institute_Emerging_green_trends_Future_of_Manufacturing_0.pdf (Accessed on 29 March 2019).
2. Beder, S., (1994). *Role of Technology in Sustainable Development*. [online] Uow.edu.au. Available at: https://www.uow.edu.au/~sharonb/RoleTech.html (Accessed on 29 March 2019).
3. Hermes, J., (2016). *How to Choose the Correct Emission Control Technology - Environmental Leader*. [online] Environmental Leader. Available at: https://www.environmentalleader.com/2016/05/how-to-choose-the-correct-emission-control-technology/ (Accessed on 29 March 2019).
4. Herrmann, C., Hauschild, M., Gutowski, T., & Lifset, R., (2014). *JIE Editorial 2014 | Life Cycle Assessment | Sustainability*. [online] Scribd. Available at: https://www.scribd.com/document/357652749/JIE-Editorial–2014 (Accessed on 29 March 2019).
5. Kim, P., (2018). *How Manufacturers are Developing and Using Renewable Energy | Smart Cities Dive*. [online] Smartcitiesdive.com. Available at: https://www.smartcitiesdive.com/ex/sustainablecitiescollective/how-manufacturers-are-developing-and-using-renewable-energy/1175001/ (Accessed on 29 March 2019).
6. *Renewable Energy in Manufacturing*, (2014). [eBook] *IRENA*. Available at: https://www.irena.org/-/media/Files/IRENA/Agency/Publication/2014/IRENA_REmap–2030-Renewable-Energy-in-Manufacturing.pdf (Accessed on 29 March 2019).
7. Visser, W., (2014). *How to Use Technology to Make Our Planet More Sustainable, Not Less*. [online] the Guardian. Available at: https://www.theguardian.com/sustainable-business/technological-innovation-sustainability-energy-green-investment (Accessed on 29 March 2019).
8. Wright, S., (2016). *1 2016 How Green Technology is Transforming Traditional Manufacturing Methods*. [Blog] apriso.com. Available at: http://www.apriso.com/blog/2016/06/how-green-technology-is-transforming-traditional-manufacturing-methods/ (Accessed on 29 March 2019).

CHAPTER 8

FUTURE ASPECTS OF SUSTAINABLE MANUFACTURING

"Surely, we have a responsibility to leave for future generations a planet that is healthy and habitable by all species."

- Sir David Attenborough

CONTENTS

The future of sustainable manufacturing (SM) will require an essential change which several manufacturing sectors are tend to enact to. In order to meet the demands of world's population, manufacturing units should engage their workforce into a more sustainable production of goods as they a play a key role in educating their employees on the SM benefits. The manufacturers should not only limit to educating their employees alone, but those beyond the company walls, such as suppliers, customers, and wider society. The major goal of SM is to produce products which are environmentally friendly while conserving the energy and resources for the future generations.

8.1. INTRODUCTION

Presently, SM is not about producing physical products alone. Due to the advancement in technologies and considering several other factors that led to a fundamental shift in the businesses due to a drastic difference product, consumer demands and the economics of supply chain. The link between the manufacturer and the consumer tend to constantly diminish as the consumer's tastes and preferences have become more personalized and customized. In addition to the diminishing link, sensors, and connectivity covert "dumb" products to "smart" products, while the products gradually become platforms and even move into the realm of services.

As the world witnesses an exponential growth in the technological sector, commercialization, barriers to entry and learning are in the verge of eradication. There are a lot of politics that prevail in the manufacturing sector in which the new entrants enter the market with new tools and tend to function only on a limited sale, which in line enables them to offer a sole province of key bureaucrats.

In the current era of SM, the larger the scale of production the more dominating it is in the market. These dominating segments would primarily have control over the value chain, innovative manufacturing models distributed small-scale local manufacturing, loosely coupled manufacturing ecosystems, and agile manufacturing which are arising to take advantage of these new opportunities. In the interim, the rising infiltrate between the product makers and the product sells has been an apparent situation presently. There has been an urge for the manufacturers to alter their SM strategies to get a better result. Manufacturers are going through the pressure and the gaining ability to rise their marketing pace and to get more involved with the customers to understand their requirements.

SM strategies witnessed a transition from stocking up products to the manufacturer on order basis due to several other factors such as to avoid the excess building and to avoid the wear and tear of stock. Due to this transition, the inventories that are with the middlemen are becoming less and less necessary.

These transitions from building of stock to build to order have made major shifts in the SM sector, and have become even more challenging to create value in traditional methods. Simultaneously, as products started to be under-valued as there are more means of access of experiences and information to those products, the values of capturing and creating has shifted from delivering physical objects to enabling that access.

Several manufacturing sectors are likely to be affected at different levels with the shifts of easy accessibility of products. In order to identify the intensity and pace of the upcoming shifts in a specific SM sector, every company should consider several factors that include the product size, complexity, sector's level of digitization, extent of regulation.

As the SM witnessed a substantial increase in the above trends, a company's key officials should try to focus primarily on the sectors which are likely to lead to consolidation and concentration and trying to evade those that are prone to fragmentation. There are three factors that drive the significant economies of scale and scope and serves as a pillar to promote a company's growth and profitability; they are:

- Infrastructure providers;
- Aggregation platforms;
- Agent businesses.

As there are numerous numbers of SM industries in the market, there tend to be a high competition among those manufacturers as they experience increasing pressure to pay attention to just one factor and leading to several distractions from becoming the top companies at the global level in its chosen role. The outcome of this could be a substantial restricting of current product manufacturers.

To further develop the potential progress of adopting a scale-and scope role, an industry should follow the leveraged developed policies, rather than focusing only on "make vs. buy" options. If the drafted policies are incorporated in an appropriate method, key players will encounter several opportunities to connect, mobilize, grow an array of new entrants and many of which will target fragmenting portions of the manufacturing value chain in order to deliver more value to their customers. The Two evolving business models, "ownership to access," and, "product to platform" seem particularly promising in terms of driving leveraged growth strategies.

As the advancement in technologies marks a key evolution in the SM industry and the advent of further comprehensive networks of fragmented and concentrated companies across an evolving range of manufacturing value chains, businesses that comprehend developing "influence points" will have a substantial strategic advantage. While the competitive pressure tends to build upon the manufacturer, the company should have a sustained focus on satisfying the customers' demands as the position of the company alone matters more than ever.

An achievement in SM is not a single path to success. There exists no master playbook to discover answers to the question of what, where, and how in a whole new market. One can better forecast the process of working of manufacturing sectors by understanding these roles, shifts, trends, and the influence points. These trends do not only act as a navigator for the existing big players in the market but also helps the new entrants the tools to successfully guide the new landscape of SM.

8.2. MANUFACTURING AND SUSTAINABILITY

According to World Commission on Environment and Development, "sustainability means meeting the needs of the present without compromising the ability of future generations to meet their own needs." In a discussion held by Holmberg (1992), Reed (1997), Harris, et al., in 2001, it is believed to have been a progressive recognition of three essential aspects of sustainable development (SD), they are (Figure 8.1):

- Economic;
- Environmental; and
- Social Aspects.

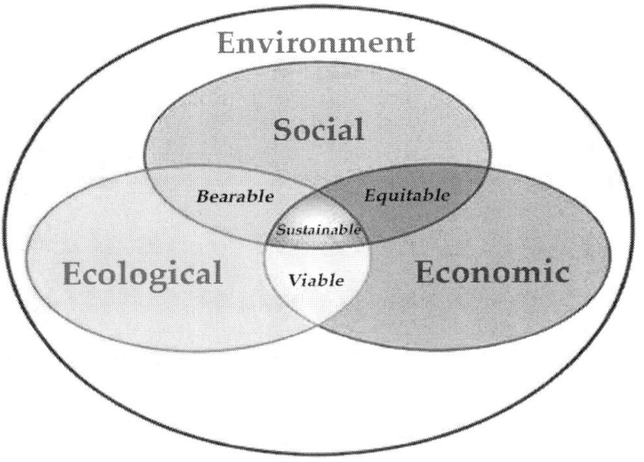

Figure 8.1: The "three pillars" of sustainability bounded by the environment.

Source: http://www.digitalxplore.org/up_proc/pdf/277-148835868921-26.pdf

Manufacturing has traditionally been associated with undesirable environmental side effects as manufacturers are responsible for the transformation of resource inputs into useful outputs (i.e., products with economic value) with limits on efficiency due to the laws of thermodynamics.

Traditionally, SM has been linked with undesirable environmental side effects as the key responsibilities tend to rely on the manufacturers in transforming the resource inputs into valuable outputs (i.e., products with economic value) with limits on efficiency due to the laws of thermodynamics.

It has been a matter of global concern and a major threat to the society for over the last four decades, that the environmental burden which are primarily associated with the industrial activities for production purposes. The key reason that endorsed the implementation of environmental degradation prevention practices is to spread awareness among the manufacturers about the impact of human activities on the global environment that could lead to a major disaster for the future generation.

These environmental degradation prevention practices have several labels for different sectors in the market such as Green Supply Chain Management, Corporate Environmental Management, Industrial Ecology, Product Life-Cycle Management, Product-Service Systems (PSS), and

Design for Environment (DFE), etc. It is also notable that there are varied elements which play a significant role in jotting the requirements that the next-generation needs in the manufacturing paradigm like environmental concerns, technology advances in modeling and simulation techniques, lack of knowledge integration and product and systems complexity.

In the recent times, the concept of SM has been developed under several labels such as; Environmentally Conscious Manufacturing or Green Manufacturing as a sub-concept of Pollution Prevention (P2).

The aim of SM is to reduce the environmental impact that is associated with manufacturing. For a long time, environmental activities have been linked with an adverse impact on the performances of any business. However, several researchers have proved the above statement to be false. Primarily, sustainable indicators reflect the association among the three aspects of substantiality and several other factors which affect them. These links are showcased in the below Figure 8.2;

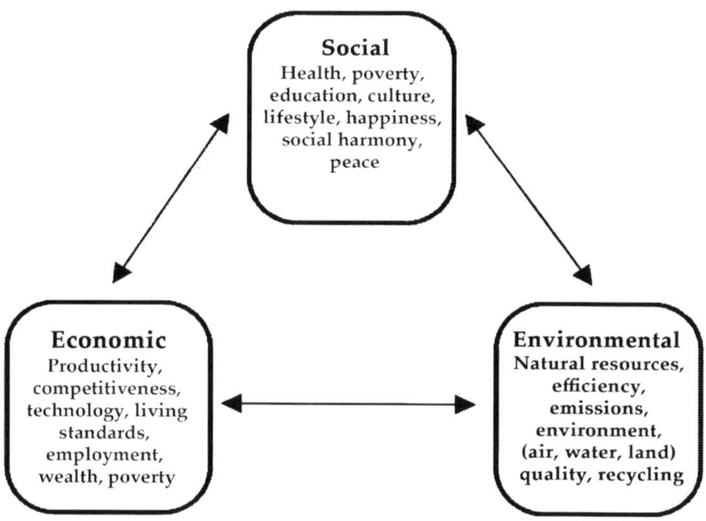

Figure 8.2: Relations between social, environmental, and economic parts of sustainability, and some of the factors that comprise them.

Source: http://www.digitalxplore.org/up_proc/pdf/277-148835868921-26.pdf

- Natural resources act as a base for various materials and for the purpose of production which in turn creates profits and job opportunities;

- Employment affects, poverty rates, living standards, and wealth creation;
- Poverty relates to social unrest and instability and crime;
- Resources like the quality of air and water will affect health;
- Resources used for production affect profits.

8.3. THE WORLD IN 2050

With the current trends in the advancement in technologies and considering several other aspects, there would be a radical rise in the world population, continued emission of greenhouse gases and the usage of materials, the entire world would witness extreme changes in the way people live and work and a complete shift in the ecosystem cycle by 2050. Certainly, as there will be an exorbitant increase in the population, enormous amount of resources and materials are likely to be consumed, and it is believed that the concomitant carbon dioxide emissions will grow under a business-as-usual trajectory, forcing the world over 2°C average warming and possibly up to 4°C – 6°C, with predicted catastrophic effects at the extreme.

A distended population results in the scarcity of several resources which people consume and the manufacturing sector is likely to encounter numerous challenges in manufacturing the products. In the course of easily-exploitable sources depletion of raw materials, the prices of physically scarce goods tend to increase to meet the expenditure on raw materials by the manufacturers. In order to meet the demands for water and land for agricultural purposes, industries will witness several new entrants that results in a battle as natural resources are prioritized to society's basic needs over industry.

The trends in the extreme weather conditions which the world is likely to witness have disastrous impacts on mankind. These extreme weather conditions may make the land inhabitable for humans for majority of the months or all throughout the year. More and more populations tend to move to the ever-growing urban cities which are less resilient to risking supply chains and unpredictable shocks. Due to insufficiency in the elasticity of urban areas, there might be an upsurge in the risk of supply chains disruption with major consequences for the manufacturers worldwide.

These projections take historical growth and efficiency trends into account and infer that business-as-usual, in this case, a drive to efficiency is a necessary, but not enough response to future sustainability challenges. Several opportunities are likely to be seen in the European and United

Kingdom market which signifies both long-term and short-term values, only if the countries adopt both, the radical innovation and the efficiency initiatives. However, implementing them is not a cakewalk. It requires legitimate policies and support from various sectors which several countries are lagging.

8.4. THE MEGA FORCES WHICH DRIVES TO THE FUTURE DIRECTION

According to KPMG International, several environmental "mega forces" such as climatic change, energy, and fuel, material resources, urbanization, water scarcity, deforestation, wealth, growth in population, decline in the ecosystem, resource scarcity and food security are believed to be the major drivers for the business change to 2035 and beyond. When the above environmental mega forces are independently studied, it is predicted to have major impact in doing a business.

However, it is challenging to comprehend the intensity of the impact as these drivers as they are irreducibly inter-linked to one another. The intensity at which these drivers influence each other are complex and takes shapes in unpredictable ways that cannot be extricated to elucidate determinate cause-and-effect relationships. This has implications in that industry will have to conceive of technologies and business proportions which meet the expectations of several limitations simultaneously.

8.4.1. Mega Force: Population Growth & Urbanization

It is estimated that there would be an addition of 3 billion along with the current world population by 2050. The research also states that population of 2050 will be richer, older, and predominantly urban, which leads to several opportunities and growth in the trade sector particularly in context of Europe and United Kingdom.

According to World Economic Forum, "Unsustainable population growth is considered to be a key driver that put societies at considerable risks." If the level of consumptions is positively linked to the growth in population, this will place an unparalleled burden on the environment and on society. This will be a situation of exogenous factor for the industries to adapt to but will provide substantial growth opportunities.

8.4.1.1. Pressures

According to United Nations 2011 reports, it is estimated that the world would have whopping population of more than 9 billion by 2050. According to the reports from Organization for Economic Co-operation and Development (OECD), African, South Asian and Middle Eastern Countries will have a foremost increase in their population.

According to the forecasts made by the World-Wide Fund (WWF), this rapid incline in the growth of population will need the worth of 2.1 planets resources to meet the requirements of more than 9 billion population by 2050. According to a report by McKinsey Global Institute, by 2030 there will also be a huge demand for water by 140% and land by 250% (Figure 8.3).

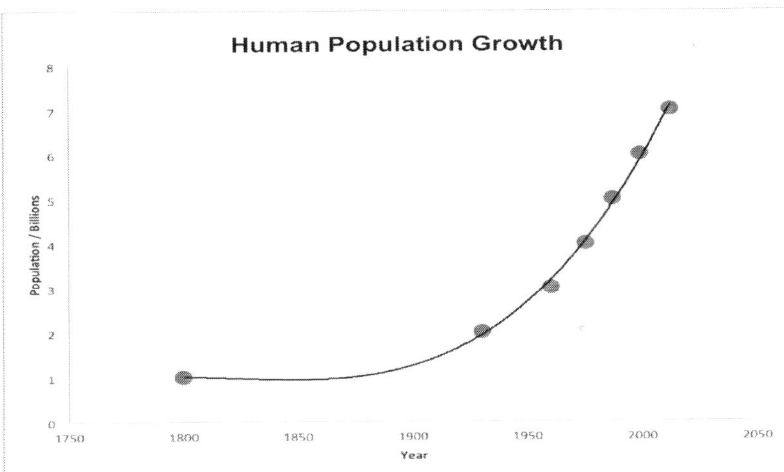

Figure 8.3: Population growth.

Source: https://upload.wikimedia.org/wikipedia/en/3/3a/Human_population_growth_from_1800_to_2000.png

Agricultural sector will have an additional demand of land for irrigation purpose such as animal products and cereal crops. The increased population will need 55% more water resource to cater the needs of people. Whereas, the manufacturing sectors will tend to lure most of the world's water which constitutes to almost 400%, while the thermal electricity will draw more than 140% and the usage of water for domestic purpose would be of more than 130%. As more water is exerted from the surface of soil, there will be a depletion of groundwater level due to which majority of world's population will live in water-stressed areas. Due to the depletion of groundwater in several areas it will act as menace to urban water supplies and agricultural lands. According to the projections made by United Nations Convention to Combat Desertification (UNCSD), it is said that by 2050 nearly 6 billion people will live in urban areas while only 0.6 billion will live in rural areas which accounts to nearly 70% of world population.

OECD projected that there will be an increase of 80% on the demands of energy and resource due to this exorbitant increase in population, with fossil fuels providing 85% of the energy mix, further aggravating the extreme climate changes due to the excess emission of greenhouse gases into the environment. Although urbanization ought to allow better management of scarce resources, an increasingly urban population will experience a warmer world with a rise in competition for food, natural resources, land, health issues, traffic congestion, and water stress.

Because of global warming and extreme weather conditions, there will be an increased vulnerability in urban areas due to concentration of assets and people, particularly infrastructure and transportation systems which are located near coastal regions. According to United Nations International Strategy for Disaster Reduction (UNISDR), the consequence of global warming and extreme weather conditions will have major impacts on the manufacturing sectors if supply chains are disrupted.

Intergovernmental Panel on Climate Change predicted that air pollution and localized urban heat waves from transport and cooling will rise the death rates particularly but not limited to people who are over 65 of years of age. With the increase in the pollution and urbanization, premature deaths are predicted to double to nearly 3.6 million deaths annually, mostly in Asian countries.

The 36 countries of OECD are predicted to have most premature deaths due to the high level of pollution subsequent to India. It is predicted that by the end of 2030 more than 33% of global population will be of over 60 years of age and the figures are likely to increase by 2050. However, the percentage of population over 60 years is currently 12% (Figure 8.4).

Figure 8.4: Urbanization.

Source: https://upload.wikimedia.org/wikipedia/commons/9/94/Mumbai_Downtown.jpg

As per Cracknell's projections, countries like United Kingdom and other European countries will have 2.8 people of working age of every pensioner by 2033 which is a decline from 3.2 in the year 2008. Countries which have higher percentage of youth will have a stronger workforce than the countries which has most people of over 60 years of age. Countries like United Kingdom have a lessening workforce which could have a greater impact on the manufactures who find it challenging to adapt. It is also predicted that Africa and the BRICS countries will primarily contribute to the World GDP, which is said to be quadruple by 2050.

Between 2010 to 2030, the world will see a growth in GDP which shall be driven by increase in machines, infrastructure, and buildings. From the above statement, it is implied that there would be a huge demand of resource and energy to meet the needs of the world.

The non-OECD countries tend to witness lesser growth but higher services in industries and manufacturing than the OECD countries which is likely to have tremendous growth in industries. The World Business Council for Sustainable Development (WBSD) predicts that the world will have 4 billion middle-income groups by 2030 which could be an unprecedented opportunity for business growth, but may come at a cost if Western consumption habits are duplicated in developing countries.

According to a report OECD, in the case where the demand for energy and material efficiency rises parallelly to the population in a business-casual set up, there is expected to be crucial impact to the entire planet which could largely decline the average consumption per capita by 14% by 2050 as it is practically impossible for the globe to sustain this whopping rise in population which consumes the same level as occurs in post-industrialized countries today.

Gross Domestic Product on a national level is calculated according to the consumption-led growth and there said to be a little appetite in the west to consume differently. Several manufacturers of European countries and the United Kingdom and for that matter, any business in general, should explore the opportunities to produce the same variety of products with less impacts and shift to different modes of consumption, which includes service provision and selling performance rather than the goods.

The most crucial aspect which is of major debate is the consumer's change in behavior and several significant programs for a change will require that social scientists be given equal footing with engineers and scientist to overcome the issue in an informed manner.

8.4.2. Mega Force: Climate Change

It has been observed that there are several reasons corresponding to Climate change. Some of those major reasons are:

- Circumstances where the weather conditions are extreme;
- Tremendous increase in the average global temperatures; and
- A troublesome situation of water availability.

But all these problems could be taken care off if proper steps are taken into consideration. This depends on few important inferences in the case of SM wherein it would be demanded that the resources must be utilized in such a way that it satisfies the basic requirements of the human population. It becomes the responsibility of the manufacturers to implement such processes that are efficient and effective and would also develop products which are environment-friendly and emit fewer amounts of greenhouse gases (Figure 8.5).

Figure 8.5: Extreme weather condition.

Source: https://upload.wikimedia.org/wikipedia/commons/9/91/Severe_weather_montage.jpg

This would help in managing the unpredictable supply chain risks in the manufacturing department or sector.

8.4.2.1. Terrible Consequences

The result of an increasing population burning fossil fuels on a business-as-usual trajectory is terrible, and it highly impacts the atmosphere around the humans. Due to continuous burning of fossil fuels, there are more chances of getting higher level of greenhouse gas emitted in the air which will continue to rise despite overwhelming scientific evidence. The evidences suggest that the inability to reduce the greenhouse gas emission would result in catastrophic consequences on the biosphere which would somehow impact the human capability to get adapted to a warmer planet (Figure 8.6).

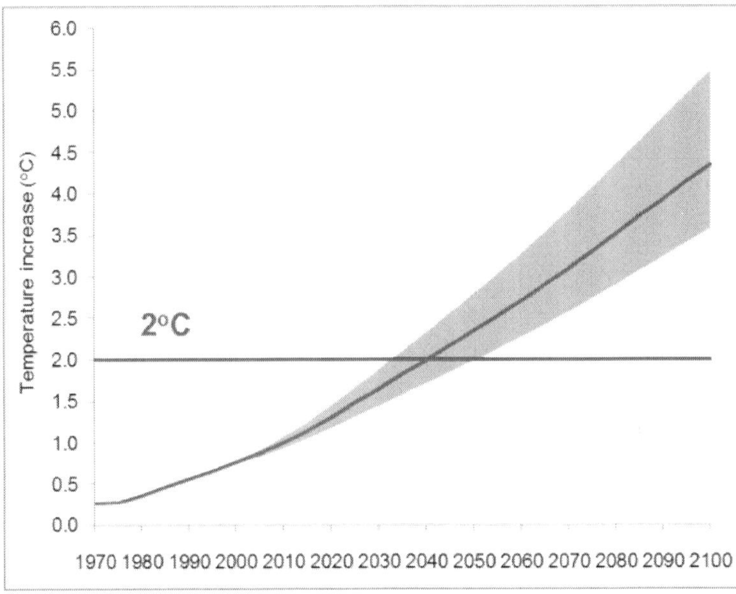

Figure 8.6: Projected future temperature increases because of increasing GHG emissions.

Source: https://assets.publishing.service.gov.uk/government/uploads/system/uploads/attachment_data/file/283909/ep35-sustainability-and-manufacturing.pdf

It is considered "extremely likely (>95%)" that it is because of various unwanted human activities that have brought such terrible consequences into existence at first place. (IPCC, 2007) According to various researchers like Cook, 2013 and others, that these human activities are the main reason behind the extraordinarily increasing amount of greenhouse gas in the atmosphere. But there are several uncertainties attached to the climate models because past predictions have been largely borne out by present evidence (Allen et al., 2013).

According to the report published by PwC, 2012, "The average annual rate of decarbonization since 2000 has been 0.8%, with carbon emission increases tightly coupled to increases in GDP." According to International

Agency, 2012, "The 2050 global energy mix is predicted to be similar to that today, and 85% of the power supply will be derived from fossil fuels."

It was believed that it is essential to meet the national greenhouse gases reduction pledges otherwise the amount of carbon dioxide that can be emitted into the atmosphere to limit warming to less than 2°C will be locked in mainly through capital expenditure by 2017. Therefore, it was important that the rate of decarbonization must be improvised from 0.7%–0.8% per annum to 5%–6% per annum by the end of 2050 which would further help in reducing 2°C warming.

This illustrates that we must consume only 1/3 of our proven reserves of fossil fuels up to 2050. If the process of decarbonizing is taken into consideration and is worked upon for twice the current average rate, there are chances that the planet would get heated up to 3°C–6°C by 2100.

World Bank in 2012 declared that various adaptation designs might fail to respond to 2°C warming. It was observed that the greenhouse gas emissions from different sectors grew at a tremendous level wherein the situation was that the growth followed a trajectory based on the historical efficiency gain (Figure 8.7).

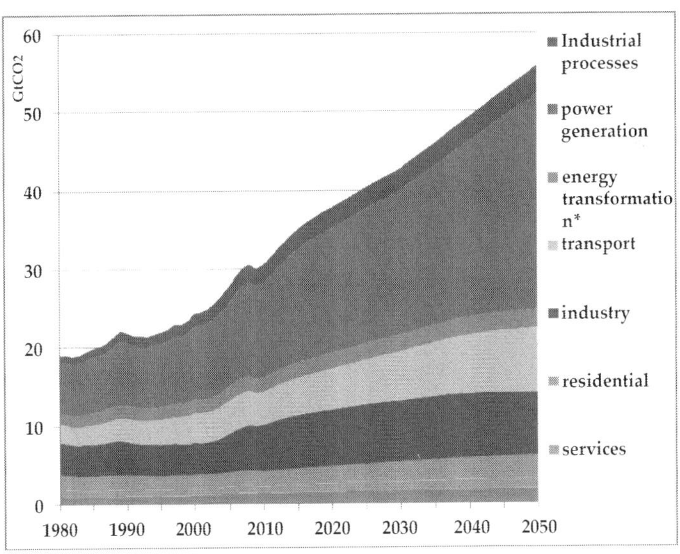

Figure 8.7: Projected GHG emissions by sector to 2050.

Source: https://assets.publishing.service.gov.uk/government/uploads/system/uploads/attachment_data/file/283909/ep35-sustainability-and-manufacturing.pdf

According to Intergovernmental Panel on Climate Change, 2007, an increase in global temperature will have impact on various things. Some of them are:

- Negative impact on agricultural yield;
- There would be a substantial increase in human deaths as there would be excess of pollution and heat in the atmosphere;
- The planet would go through a shift in the weather patterns;
- There would be more frequent and extreme weather events;
- There are chances of regular occurrence of natural disasters like excessive heat waves, droughts, and floods;
- Increase in ocean acidification;
- A tremendous level of land-use change; and
- Huge loss of biodiversity.

Global warming results in increasing the level of urban air pollution. It could be framed as a major cause for premature mortality. Like it has been discussed above that huge biodiversity loss is expected due to global warming, wherein biodiversity comprises species and primary forests loss. It has been projected that by the end of 2050 there would be a decline of almost 10–15 percent of various important elements of biodiversity including human lives.

Consequently, this would have huge negative impact over those elements which are more likely dependent on the land for their sustenance and would also impact the agricultural supply chain in the future. For example, the lost value associated with deforestation is estimated to be between US$2trn – US$5trn per year (OECD, 2012).

8.4.3. Mega Force: Material Security

For manufacturing, a product, not one kind of raw material is used. There are several kinds of raw materials which could be an ore, few minerals or fossil deposits. These products might not be present in the future for manufacturing purpose. It has been believed that the process required for the extraction of these raw materials would not be economical. This is so because they could be very easily exploited and high-quality sources are depleted. Therefore, to manage the supply chain disruption, it is now the manufacturers responsibility to come up with few material-efficient processes which would help them survive the market competition.

8.4.3.1. Pressures

In the past few decades, a huge rise has been witnessed for different engineering materials and various other rare earth elements at a global level. It was in the year 2012; Chatham House forecasted that "Demand for metals will increase by 30%–50% by 2020 and by up to 90% for steel and 60% for copper by 2030, compared to 2010."

Depending on the forecast, it was also believed that there would be tremendous growth in the worldwide population and growth in wealth demand would also stimulate which could be at least double based on current levels by 2050 (Allwood et al., 2011). Chatham House, 2012, also stated that. "China will consume up to 50% of global metals in 2020, much of this coming from imports."

In the case of renewable energy, rare earth elements play an important role and are used for clean efficiency technologies, including photovoltaic films, magnets used in wind turbines, batteries used in electric vehicles and phosphors used in low energy lighting. These materials are quite high at risk depending upon the number of criteria and scarcity. Their criticality is also dependent upon production concentration, reserve distribution, recycling rate, substitutability, and governance.

A lot of pressure has been put in the materials due to various factors. The top factors are:

- The depletion of economic reserves of materials;
- An environmental and monetary cost associated with the exploitation of lower grade minerals.

These factors force the people to adopt various conservation and substitution measures to help restore the rare materials on the planet (Allwood et al., 2011). World Economic Forum, 2012b illustrated that "A failure to respond to "peak metal" could put US$2trillion of economic output at risk in 2030." With the increase in globalization, it is quite evident that there is going to be some tough competitions for resources and people would go to any extent to protect the resources.

This is so because the countries that already are exporting rare earth materials would now fight for their own survival first, that means they would first keep or hoard these materials for their own use and then they would prefer exporting it to the preferred countries.

It is also estimated that a "peak oil" scenario might occur before the end of 2030 and this would consequently reduce supply of conventionally extracted oil. This would also result in increasing hurdles in the way of extracting materials and discovering those materials at the first place. Situations like these might result in the increase in the price of oil-based commodities and suppressing demand (Allwood et al., 2011). Materials based on renewable feedstocks, including wood, paper, and bio-plastics; rely on availability of favorable land.

The trends in competition for land to 2050 show an increase in land for bio-energy, crops, and livestock, with forest and other land decreasing. The exceptions are scenarios implementing a carbon tax and a lower meat diet where more land is converted back to unmanaged forest (Smith et al., 2010). The opportunity to significantly expand renewable materials production is therefore considered to be limited, which impacts the number of biofuels that can be produced and thus the penetration of less carbon-intensive fuels as a strategy for decarbonization of the energy supply.

After the extraction process has been conducted, the by-products could be harmful for the local atmosphere. This situation could arise in the case when proper precautions have not been taken by the concerned authorities due to which waste impacts may grow as ores are depleted (Norgate et al., 2007). This would generally demand for remediation to restore local eco-systems. Also, it has been observed that well-run mines can be beneficial to local communities and ecosystems providing employment and improving biodiversity (IIED, 2002).

8.4.4. Mega Force: Water

As more and more water will be consumed by people, the possibility of scarcity of water resource is likely to occur. The usage of water is meticulously linked to food production, environmental integrity, and energy. The future struggle of this finite resource may urge the manufacturers to have a straight competition with the food production and the basic needs of society along with subsequent political allegations. Researchers also suggest that the industries should make less use of the fresh water which is available to them in order to mitigate against this possibility.

8.4.4.1. Pressures

The key factors which results in the major demand and greater levels of consumption of water by 2050 are;

- Population growth;

- Urbanization;
- Demographic shifts.

As projected by OECD, the demand for water at a global level will rise by 55% by 2050 in comparison to the year 2000. With a 400% increased demand from manufacturing and 140% from electricity generation.

According to World Economic Forum, the chances of water supply crises over the next couple of decades are very high which approximately constitutes to an average of 3.8/5.0. This water supply crisis results in the depletion of groundwater which impacts the energy and food manufacturing sectors to a great extent (Figure 8.8).

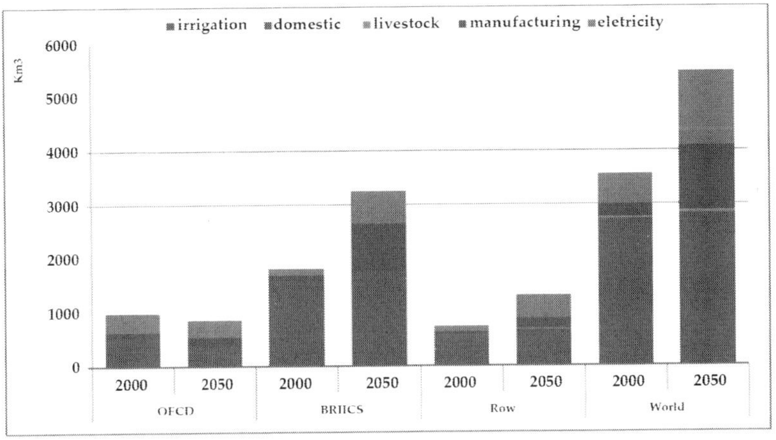

Figure 8.8: Bluewater demand to 2050 by sector.

Source:https://assets.publishing.service.gov.uk/government/uploads/system/uploads/attachment_data/file/283909/ep35-sustainability-and-manufacturing.pdf

According to a report by the Royal Society, more than 70% of water at a global level pulled out for agricultural purpose whereas 19% for the purpose of industrial sectors. The European countries and the United Kingdom's public water supply accounts to 30% of the total, industries by 9% and electricity supply by 55% whereas, 90% of water is being abstracted in India for the purpose of agriculture. The consumption of water in the United Kingdom is greatly dependent on "virtual water" which are being imported and embedded in goods that are produced from water drawn in their countries of origin.

According to Intercontinental Exchange report of 2012, of the total consumptive water used in the United Kingdom, three-quarters of the total is met by water resources from other nations. Energy is used for the purpose of cleaning and movement of water. As estimated by the Royal Academy of Engineering, in California almost 19% of electricity and 33% natural gas is primarily used in water-related activities.

According to McKinsey Global Institute report, due to high rate of population, more groundwater is required to meet the ends of people, leading to the depletion of groundwater in a very short time. Due to this depletion, more energy will be required to extract the same amount of water, which in turn has a huge impact on greenhouse gas emissions, if the required energy is produced using fossil fuels. Likewise, the water requirements for fuels can be substantial, with biodiesel having a water footprint 60,000 times greater than the fossil fuel equivalent.

According to United Nations, Educational, Scientific, and Cultural Organization (UNESCO) the middle-class and the upper-class people tend have constant variations in the consumption of food where more of meat and vegetable oils are consumed instead of pulses and grains which add more pressure on water supply. For instance: One kilogram of beef requires 15,500 liters of water to produce, while the equivalent amount of wheat requires only 1,300 liters.

According to the OECD, by the end of 2030, more people will start living in urban areas than the rural areas. To be precise, more than 60% of the global population will live in cities than the rural areas. Though this increase in urbanizations has a positive impact in managing the water easier, it also places growing pressure on local water resources, which is further aggravated by climatic changes. This depletion of water resources also affects the trends of the hydrological cycle in uncertain ways at a global level.

8.4.4.2. Extreme Weather Events

As the average global temperature increases, by the end of 21st century, it is likely predicted that it will have a major impact on frequency of weather patterns and severity of extreme events. Global warming also increases the threat of damage particularly in the coastal region's power supplies, infrastructure.

According to UNISDR, "the global nature of production means that disruptions in other parts of the globe can significantly affect the supply chain." An example to the above statement is the 2011 monsoon that hit Thailand's more than 1000 factories which were the supplier of 804 global companies.

There were several industries which were affected by this monsoon hit, particularly the computer hard-drive supply chains, which declined the computer manufacturers income between 5–10% by PwC. Parallelly, the supply chain of Toyota and Nissan companies had a major impact due to which the companies had to suspend the manufacturing of car thereafter, costing them US$1.25bn and US$1.4bn respectively.

While the majorly affected countries due to water crisis will be in the Asian countries, some parts of United Kingdom are likely to witness droughts, water stress and heavier floods which affect several elements such as transportation, infrastructure, transmission, Information, and communication technology and agriculture.

Taking into consideration the two aspects such as growing population and extreme weather conditions, the manufacturing sectors in the United Kingdom region should take necessary measures for the potential for some disruption along its entire value chain. These measures may need the advancement in reconfiguration of supply chains and ultra-efficient processes for resilience rather than efficiency.

8.4.5. Mega Forces and Supply Chain Vulnerability

A great threat which rises the vulnerability of the supply chain are the mega-forces. These increasing mega forces like the extreme weather conditions, global warming, and other events which damage the environment on a whole will end up in the volatility of excess usage of land, raw materials or disruption of crucial infrastructure, particularly in relation to the operations that takes place in the coastal areas.

Additionally, yet other element which contributes to the vulnerability of the supply chain is the globalization. The key priorities of businesses are the short-term cost efficiency, outsourcing, and offshoring manufacturing; reduced inventory capacities tend to rely on single sourcing rather than having alternative suppliers from which to draw from. The chief impacts of disruption which are reported by the manufacturers are the loss of orders and revenues, followed by a delayed cash flow.

To see sustained revenue despite the cost and lack of capability, the manufacturers of United Kingdom have two key practices such as:

- Have a part of their supply chain overseas; and
- 20% or half of the suppliers will be from outside of the country with Asia being the biggest supplier over half of the manufacturers.

According to a report by Deloitte, the Pacific and Asian countries supply critical materials to the United Kingdom's manufacturing sector, which includes copper, aluminum, iron ores, and industrial minerals.

According to EEF, (which is an organization for manufacturers in the United Kingdom) in order to overcome the supply chain vulnerability challenge, the United Kingdom manufacturers are required to take three measures such as: should start increasing the collaborations with several nations; the sourcing of various suppliers across the world, despite the increased transaction costs and quality control issues; and on-shoring and increasing their use of local suppliers.

According to the UNISDR, even though the manufacturers take up various steps to tackle supply chain vulnerability, most of the business particularly the Small-Scale Enterprises tends to fail in drafting an appropriate plan in managing the disruption.

According to Deloitte, "a resilient supply chain will have a number of characteristics that allow it to adapt to disturbances with minimal financial impact." Redundancy is another significant feature is having an alternative supply chains which also includes the alternative mode of transports. However, redundancy is ostensibly at odds with efficiency, where buffering capacity is removed to save costs.

8.5. THE FUTURE TRENDS OF SUSTAINABLE MANUFACTURING (SM) IN THE UNITED KINGDOM

8.5.1. Volatility of Supply

The primary reason for the rapid rise in urbanization is the growing population worldwide. It is estimated that more than 70 percent of people will migrate from towns and villages to the cities by 2050 thus increasing the demands various resources like the raw materials, energy, and water. The outcome of increasing demand of these raw materials will be; high level of competition among the manufacturers with potential disruptions in their supply. Typically, the impact of competition will result in high prices of products and they tend to become more unstable. Those countries and companies which absorb the ways to manufacture their products with less of these inputs will be more resilient to these effects.

8.5.2. Climate Change and the Increased Vulnerability of Global Supply Chains

By 2050, the differing trends in the climate will have a significant impact at a global level particularly for the manufacturing sectors. The series of impacts include extreme weather conditions and a rise in the sea level. Manufactures in the United Kingdom will encounter disruptions not only in their domestic supply chains but in their global supply chains as well.

8.5.3. Greater Use of Regulation, Potential 'Pricing of the Environment'

The appropriate usage of regulations is likely to concentrate on increased promotion of resource productivity. The recent European Union (EU) legislation is an instance, where it aims to divert electrical equipment waste away from landfill. National and international responses are likely to include tougher environmental standards for products and new ways to price natural resources and ecosystem services (Figure 8.9).

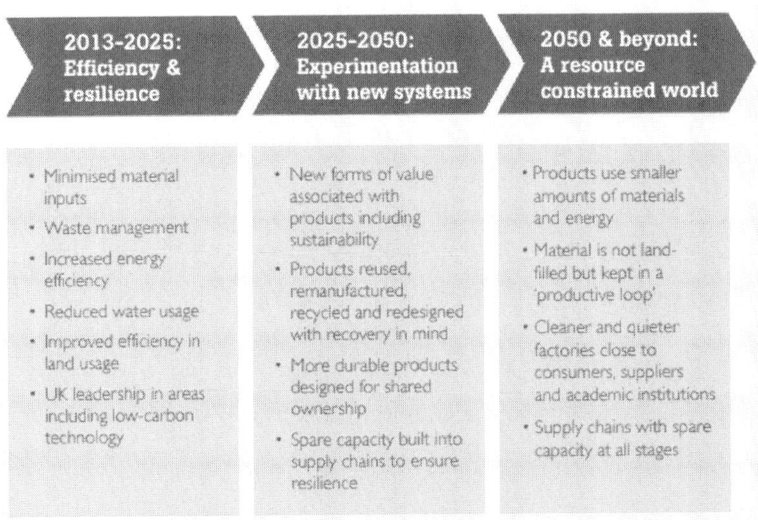

Figure 8.9: Three phases in the shift to sustainable manufacturing.

Source: https://www.ifm.eng.cam.ac.uk/uploads/Resources/Future_of_Manufacturing_Report.pdf

8.5.4. Energy Intensive Industries

The term energy-intensive industries include all those industries which consumers of enormous amount of energy or generate greenhouse gases which is an inherent part of their manufacturing process. Several industries can be given as an example which consumes enormous amount of energy, they are;

- Chemical industry;
- Iron and steel industry;
- Nitrogen fertilizers;
- Cement industry;
- Paper and ceramics.

According to a report published by Centre for Low Carbon Futures, all the above industries are said to emit more than 45% of carbon dioxide from both the public sectors and the United Kingdom's businesses, which generates a whopping £5billion GVA with 125,000 workforces.

To mitigate the GHG emissions and to keep a check on the rising fuel prices, energy efficiency programs have been extensively implemented across the regions of United Kingdom. The outcome of these extensive programs has revealed tremendous results through which the industries started to adopt the latest technologies while effectively managing the resources. According to Bassi et al., "Under a 'business-as-usual' scenario, projecting historical efficiency trends, the industries will decline as the cost of fuel rises further."

The industries should study and develop those technologies which are required to meet the goals of 2050. The industries should also make sure that the technologies that are developed meets the criteria of reduction in wastage of raw materials and resources, and is supposed to deploy the best and most effective practices and ensure the effective usage of under-utilized resources like;

- Recycling waste heat;
- Innovation in new materials and processes;
- Developing cross-sector synergies;
- Use of decarbonized electricity; and
- Development of carbon capture and storage systems.

In order to achieve the goal of effectively managing the raw materials and resources, industries should increase the Research and Development activities.

According to Climate Action Network Europe, several research and development of new technologies are reported to be introduced in the market such as cement kiln processes, clinker replacement, and Ultra Low CO2 Steel, in order to combat wastage of resource. However, in order to reach the desired target, it is required to deploy in conjunction with carbon capture and storage.

8.6. BUILDING BLOCKS OF THE FUTURE OF SUSTAINABLE MANUFACTURING ENTERPRISES

As believed by the researchers, it is far too early to a task to predict what 4.0 industrial eras will look like as each of the earlier industrial revolution took a minimum of 80 years to complete. However, in order to have a brief idea on 4.0 and with the experience of industrial evolution, the following points will contribute to the success of any manufacturing enterprises:

(1) **Digital twins and digital threads:** The concepts of digital twins and digital threads are studied under the digital manufacturing techniques which were primarily spearheaded only by the defense communities. Several manufacturers across the globe started using the smart manufacturing/ digital manufacturing in their day-to-day life, and the concept of digital twins and digital threads gave the manufactures a better insight to improve their future programs.

The digital twins refer to a physical asset which includes design specification and engineering models that describes its geometry, materials, components, and behavior. On the other hand, the digital threads refer to the communication framework which links the data from various areas of the assets and provide an integrated view of the asset's data throughout its lifecycle across traditionally siloed functional perspectives. The key characteristic of digital thread is what the manufacturing is intended to achieve by offering the "right components to the right place at the right time."

(2) **Total supply chain transparency and visibility:** The manufacturing sectors are gradually accepting the concept of system engineering. The concept of system engineering optimizes the manufacturing operation (i.e.,) from start to finish, more

effective, which includes the location of production, distribution center, to advocate the optimizing of the entire supply chain or the entire supplier base. With the advent of technology, the industrial sector is a step closer to the whole supply chain visibility and transparency.

(3) **Innovative materials:** From the Stone Age era to the current period, majority of human evolution was involved in the innovating and creating new materials which could serve the purpose of requirements such as stone, bronze, iron, and so on.

The topic of material innovation would always remain a significant subject in the field of technology, science, and financial policies. Thus, materials and manufacturing are intertwined with each other and can be inseparable. Materials provide the "work" for manufacturing, whereas manufacturing adds value to raw materials.

According to experts, the timeframe that takes for a material to mature from the laboratory bench top to a market place is between 20–25 years. An industry should make sure to accelerate the process of research of a product, develop them and deploy the final products to the market. The industries should also take into consideration the integrated occupational materials engineering and advanced manufacturing techniques.

(4) **Advanced metrology:** As a general rule of thumb, the accuracy of measuring should be at least an order of magnitude higher than what it is meant to measure. As Nano-processing is becoming a routine technique in certain sectors, the demand for more advanced metrology will rise accordingly.

(5) **A skilled workforce for intelligent manufacturing:** To achieve the desired goal by 2050, every industry should possess a skilled workforce who can contribute to the innovative manufacturing techniques. The biggest threat to the next -generation intelligent manufacturing sector is the lack of skills workforce worldwide. This lack of skilled workforce majorly comes from two proportions;

Primarily, as there is an increasing number of under-employments worldwide, there are not enough people to meet the required skill set to fill the positions in the industrial sectors; which means that there is a high level of insufficiency in the quality and quantity of the workforce who involve in the future intelligent manufacturing. It is predicted by the experts that the United States will lack 3.5 million manufacturing operators by 2030. Secondly, even though if the apprenticeship has been recognized as a good method for emerging a skilled workforce, most apprenticeship programs are merely not scalable.

8.7. CONCLUSION

Thus, in the above chapter, one can get a detailed knowledge of world in 2050 and the key changes that are believed to be probable such as social changes and biophysical changes. Currently, there are several challenges that manufacturing industries are familiar with. But, for any manufacturing sectors to achieve sustainability, there are several crucial issues which must be primarily dealt with.

From the data combined from case studies of companies and from various industrial reports, it is interpreted that the manufacturing sector will predominantly stress on material and process efficiency in lieu to have a world-wide solution on complete manufacturing sustainability as there will be financial and resilience drivers to do so. At this juncture, the consequences of population growth and climatic change will start becoming evident and thus will not drive substantial change beyond "eco-efficiency" until post-2030.

Further out to 2050 and beyond the manufacturing industry will need to develop novel innovations and methods of production in order to meet demanding sustainability targets and create new economic growth.

It is going to be a rather different world by 2050 than what we see today. The world is changing in such a pace that it is expected to be constrained in a radically different way by 2050. The geopolitical and the man-made effects of growth in worldwide population and the climate change is expected to reframe the complete manufacturing sector, including several other sectors which are in line with manufacturing. It is also evident that the manufacturing sector will have to capitalize upon through which it can contribute to the economic growth of the countries.

The outcome of these various opportunities which are expected to be provided by the manufacturing are nothing but the consequences of increased population and the climatic change which also includes the inter-dependent risk from global warming (i.e.,) increasing global temperatures, regional water scarcity, mass urbanization, material resource scarcity, and demographic shifts. For any manufacturers to overcome these challenges, they should explore the ways by which traditional methods of manufacturing can be conceived in several new ways, moving away from primarily focusing on process of optimization. It is unlikely that the manufacturing sector can do this alone and will require support from all societal actors.

8.7.1. Drivers to 2020/2030

As per research evidences, it is predicted that the resource efficiency programs and the energy efficiency programs will rule the manufacturing sectors by 2030. The reason as to why the efficiency programs will dominate is to enlighten the manufactures the criticality of the current scenario of depleting resources worldwide, and the increasing and volatile commodity prices and the probable of geopolitical resource protectionism of important metals which will reinforce the requirement to decrease the actual prices of energy usage and raw materials.

The sinking capital costs in infrastructure and industrial design is said to pressure the physical variations which can be expected to take place, and by which these variations are said to provide key opportunities to the manufacturers to capture value. The worldwide market for the efficiency of several resources is predicted to be US$1.2trillion by 2020, while the European countries and United Kingdom are said to capture value in the efficiency in SM, building, and transport efficiency sectors including the water and fuel treatment technologies.

To better tackle the waste management and to offer extra revenue to the manufacturing sectors, the companies must opt for creating values for the current under-utilized products including the low-grade heat. In order to add more value to the economy, apart from effectively managing the waste, the company should also back-up the intellectual property in the form of business model and technological innovation, which aids in adding more revenue as well as increases the per capita income of the nation.

The programs which give crucial knowledge to effectively decrease the energy and resources used are the Industrial efficiency programs. However, a main challenge to an important development is that economic growth is currently predicated on rising consumption. Early wins will be in the B2B and M2M sectors as organizations hedge against the rising prices of resource and energy usage.

8.7.2. Drivers to 2050

Due to the substantial increase in population, which is estimated to be 10 billion by 2050, numerous numbers of resources are required to meet the needs of the people. Parallelly, the world should also be ready to encounter the sustainability challenges which will be occurred due to the growing population as there will be a major scarce of key resources and abundant labor. Meeting the sustainability challenge refers simultaneously address the use of water, resources, and greenhouse gas emissions which are primarily influenced by the growing population.

As of reports of several sectors regarding the sustainable targets to 2050, it is believed that there could be insufficient resources to meet the challenges of sustainability. Historical levels of advancement towards single-issue targets such as decarbonization are insufficient for this mission to be accomplished.

To successfully develop suitable technology to meet those encounters necessitates radical innovation that will take years or decades to research and develop and those developments are to be scaled from now. Currently, the policies that support this effort are required to work on a faster pace to overcome the drastic changes at a global level by 2050

Due to an increase in population and in turn scarcity of resources, there could be a scenario where the manufacturers must compete to seize their intended resources, which are of basic needs of society. Thus, the manufacturing sector and their supply chain should think of more creative components to their products, and the contemporary idea of consumption and economic growth will need to be negotiated.

The contemporary idea of consumption and economic growth would necessitate the reconfiguration of the manufacturing sector, including optimizing for the size and geographical location of factories and supply chains as well as reimagining the role of manufacturing in the context of society so that social opportunities can be captured alongside economic value.

REFERENCES

1. Hagel, J., Kulasooriya, D., & Seely, B. J., (2018). *The Future of Manufacturing Making Things in a Changing World.* [online] Available at: https://www2.deloitte.com/insights/us/en/industry/manufacturing/future-of-manufacturing-industry.html (Accessed on 29 March 2019).
2. Pathak, P. M. P. S., & Sharma, P., (2018). *Sustainable Manufacturing: An Innovation and Need for Future.* [eBook] pp. 21, 22. Available at: http://www.digitalxplore.org/up_proc/pdf/277–148835868921–26.pdf (Accessed on 29 March 2019).
3. Tennant, M., (2018). *Sustainability and Manufacturing.* [eBook] pp. 6–8. Available at: https://assets.publishing.service.gov.uk/government/uploads/system/uploads/attachment_data/file/283909/ep35-sustainability-and-manufacturing.pdf (Accessed on 29 March 2019).
4. *The Future of Manufacturing" A New Era of Opportunity and Challenge for the UK*, (2018). [eBook] pp. 18–20. Available at: https://www.ifm.eng.cam.ac.uk/uploads/Resources/Future_of_Manufacturing_Report.pdf (Accessed on 29 March 2019).
5. Wang, B., (2018). The future of manufacturing: A new perspective. Volume 4, Issue 5, *Engineering, 722-728.*

INDEX

N

O

P

Q

R